Transforming Organizations in Disruptive Environments

Igor Titus Hawryszkiewycz

# Transforming Organizations in Disruptive Environments

## A Primer on Design and Innovation

palgrave
macmillan

Igor Titus Hawryszkiewycz
Faculty of Engineering and Information Technology
University of Technology Sydney
NSW
Australia

ISBN 978-981-16-1452-1 ISBN 978-981-16-1453-8 (eBook)
https://doi.org/10.1007/978-981-16-1453-8



This Palgrave Macmillan imprint is published by the registered company Springer Nature Singapore Pte Ltd.
The registered company address is: 152 Beach Road, #21-01/04 Gateway East, Singapore 189721, Singapore

To my grandchildren Isabella and Michael.

# Preface

This book is written for those who want to find out about transforming organizations and their businesses or participating in a transformation. Transformation is a complex process that involves many areas. These include technology, business and business process, social and organizational structures, as well as environmental issues such as sustainability and disruption. It is of particular interest to those involved in designing new processes that rely on technology to deliver value to businesses and communities, and their stakeholders in disruptive environments.

This book introduces the areas involved in transformation in an introductory way. It describes how they must be addressed during a transformation. It does so in a structured way following a process practised in many industries.

The book is suitable for both beginners in transformation and practitioners interested in innovative ways to transform systems. To cater for different interests, the book describes transformation in four parts. Each part can be used independently of others. For those who only want to know the general background can read Parts I and IV. Part II can then be read next to find out about ways to manage knowledge. Those, who are now involved in a transformation process can read Part III, which deals with identifying real problems and ways to solve them.

In Part III, the book shows how methods and tools that have been learnt during globalization, especially on ways to be creative and innovative, and to collaborate, apply to dealing with disruptions in the increasingly fragile world faced with challenges such as disruptions caused recently by the COVID-19 pandemic and those expected from climate change.

## Part I – Setting the Background – Where Is Transformation Happening?

The first two chapters set the background and introduce readers to today's business environment. The first chapter introduces today's environment and the way organizations work. The second chapter provides a systematic approach to describing complexity in this environment and ways to create solutions. It emphasizes the need to be creative as you learn about an organization and deliver solutions. The importance of teamwork and collaborative practices is stressed.

## Part II – Values to Be Achieved in Any Transformation

In this part, the book describes transformation as not only technical in nature; it also includes changes in the way people work and services they need. ► Chapters 3 and 4 describe people values and organizational values and their importance in transformations in disruptive environments. The book sees values as critical and stresses value as one of the most important issues in transformation; the new system must not only be technically sound but satisfy people's needs. ▶ Chapter 3 describes personal values, followed by ▶ Chap. 4 that describes values of businesses and cities. It describes how beliefs and values can influence people's contribution to transformation and to work together to improve business and community development.

▶ Chapter 5 provides a framework for ways in which knowledge is developed during a transformation – how to organize the collected views of stakeholders into a form suitable for designing the transformation. Knowledge itself is central to transformation as knowing what is happening, knowing what is needed and knowing how to provide it are all central to an effective transformation.

## Part III – Identifying and Solving Problems

Part III begins with ▶ Chap. 6 by developing a model that puts the collected data into a structured visual form suitable for decision-making. It emphasizes the importance of visualization in developing an agreed description of where an organization is now.

▶ Chapter 7 uses knowledge gathered about the organization and its values to identify what a transformation should achieve. It compares what is happening in the organization to best practices found in industry and disruptions like COVID-19 in the environment. It then uses values of affected stakeholders to evaluate and identify issues in the current system. This includes analysis of causes of the issues using tools such as fishbone diagrams to identify problems and rank them for action.

▶ Chapter 8 then provides a systematic process to transform businesses to address the issues and problems identified in ▶ Chap. 7. From an information technology perspective, it could be argued that recently businesses have become global, usually facilitated by innovations in technology. The focus is on technological innovation and using information to respond. In this period, businesses have focused on creativity and innovation to develop and maintain competitive advantage.

## Part IV – Transforming in Disruptive Global Environments

Part IV addresses the challenge of creating transformations that mitigate to reduce damage from any disruptive cause by transforming the organization in general through three well-documented stages : mitigation or being ready for disruption, response during the disruption and recovery following. ▶ Chapter 9 covers major disruptions and ways organizations respond to them, which include mitigation, response and recovery. Then, ▶ Chap. 10 describes the how to mitigate for potential disruption caused by climate change, seeing it as part of the triple bottom line in the transformation. ▶ Chapter 11 outlines how cities participate in transformations by working with businesses to create value for both, and ▶ Chap. 12 then concludes with some speculation on future developments during and following the COVID-19 pandemic.

## What Value Will You Get from the Book?

Transformation has of course been a historical phenomenon over time. Building of railways, cars and others have transformed societies over centuries. Such transformations have often been on a global scale, which itself has set the environment that called for changes to the way societies and organizations operate.

The book does not focus on any disruption but approaches disruption from the general sense It views disruption from the perspective of value – how does a disruption effect the values of people and businesses. These are common to any disruption. In wildfires, we mitigate by making breaks between our house and clearing growth around the houses. Then we respond by trying to put out fires and recover by rebuilding.

The focus generally now is to deal with disruption. There are also increasing natural disruptions, caused by climate change, which similarly call for businesses to respond. From an implementation viewpoint, transformation in this book takes place in a few stages.

## Using in Educational Settings

The four parts of the book follow a roadmap that starts from a discussion of the environment, the problems caused by disruptions, followed by finding problems and then solving them.

Each chapter starts with a set of learning objectives and is followed by questions related to case studies. As students progress through the book, they are challenged by questions relevant to the chapter being read, which follows a road map that starts with general description, then builds models, identifies problems and creates solutions.

Each chapter is followed by a set of readings.

## A Focus on Design Tools and Methods?

The book uses methods known as design thinking. These have been successfully used in the past to create innovative solutions – and this book shows how to use them to address today's transformations. It also integrates them with a knowledge framework. The book can be used simply to read and develop knowledge about what is important in transforming organization through creative use of technology to create value for stakeholders.

**Igor Titus Hawryszkiewycz**
Mosman, NSW, Australia

# Acknowledgement

Special thanks are made to Matthew Bramich and Nicholas Bramich for their careful reading of earlier text versions, and many contributions leading to improved clarity of complex issues especially in Part II, and contribution to clarifying the transformation process in Part IV.

# Contents

# About the Author

**Iqor Titus Hawryszkiewycz**
has extensive knowledge about business system design developed over several years. He has developed methods for database design, structured systems design, collaboration, and, recently, design and transformation in complex and disruptive environments. As part of his research, Igor has developed a process for designing systems in disruptive environments that call for continuous innovation now increasingly needed in business and society.

His research has always emphasized application of research results while consulting to several government departments and industries, as well as developing successful public short courses at Universities.

He has over 300 hundred research publications and 6 textbooks, one in second and another in their fifth editions.

# Setting the Background

Contents

# Organizations in a Complex World

## Contents




I. T. Hawryszkiewycz, *Transforming Organizations in Disruptive Environments*,
https://doi.org/10.1007/978-981-16-1453-8_1

This chapter provides an overview of the environment in which transformation takes place. It also describes what organizations and businesses are and how they are changing and introduces the reader to how organizations operate in an increasingly dynamic and complex environment. Increasingly, change in such environments often results from disruption, which may be local to a business or industry.

This chapter then describes the importance of transformation to create new, innovative ways of working that deliver way throughout the organization.

**Learning Objectives**

- What are organizations?
- Businesses as part of an organization
- Why is change and transformation happening?
- How is disruption affecting people and their values?
- How to design transformations to create value
- The role of technology

## 1.1 Introduction

An increasingly changing world requires organizations themselves to change. The word "transformation" is now increasingly used to describe organizational change. The goal of transformation is to develop ways for enterprises to work in better ways to create value for their people. Transformation requires careful design to bring in changes that create value while working in complex environments. Terminology becomes important when describing transformations in complex environments. One way to see today's world is as an ecosystem of people, and businesses connected into organizations as shown in ◘ Fig. 1.1. The term "organization" is sometimes hard to define as it is often used interchangeably with other terms. For example, any large business is often referred to as an organization. A global supply chain is an organiza-

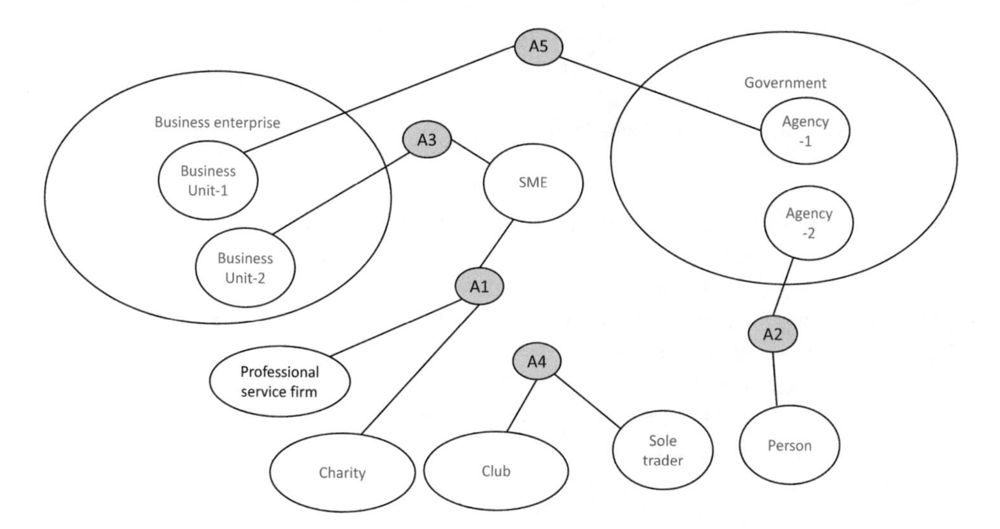


◘ **Fig. 1.1** The environment and its organizations

tion composed of many businesses. Each such business is managed to achieve goals by acquiring resources and using them to create products and services for other businesses.

This chapter introduces the terminology used in this book. ▶ Chapter 2 then describes the resulting complexity and its effect on transformation. This book sees organizations as collections of businesses in an increasingly complex world (Rowntree et al., 2017). Each business in an organization joins the organization to satisfy its values, contribute to the values of other businesses, and must run in a commercially successful way.

The term *business* is now used freely to mean more than commercial activity. Many governments, or even charities, see themselves as running a business. *Generally, business is increasingly used to mean that the organization uses funds wisely to deliver products or services to others.* Thus, delivering food to the homeless can be a business—how do you acquire the food and deliver it to needy people. Providing financial advice is equally a business but now the outcome is financial. An organization can then be defined as a set of connected businesses. Each business uses its funds to achieve organizational goals.

Businesses acquire resources from other businesses in the organization and use these resources to provide products and services to other businesses in the organization.

What is a business? There are many kinds of businesses such as:

- A global business such as today's technology companies Apple, Microsoft, or Amazon
- A business unit within a large business enterprise
- Professional service firms some themselves global but with units in different regions
- A sole trader that provides services to other businesses
- A club of people with common interests
- A small to medium enterprises (SMEs)
- A charity
- An individual such as a small trader.

Businesses connect with other businesses into business networks to carry out activities that generate value for each of them. A business network (Furr et al., 2016) is itself an organization. In the past, the term "business network" usually referred to a network of businesses whose goal is to create commercial profit. Now many other kinds of entities are businesses, and the term "organization" includes any kind of business. For example, ◘ Fig. 1.1 illustrates a few kinds of organizations.

A1. A charity supported by a professional service firm to run its accounts and an SME to manage its services.
A2. An individual seeking support from a government agency.
A3. An SME using the services of a business unit of a large enterprise.

A4. A sole trader working for a club.
A5. A business unit of large enterprise providing services to an SME.
A6. A business unit of a large enterprise providing services to a government agency

Organizations are now more dynamic as businesses leave and join organizations (Furr et al., 2016). Each such business has both its social and commercial values. Its social value is to provide services and products to other businesses, while being commercially viable to survive. Organizations often tend to evolve in ways that creates value for each business in the organization.

### 1.1.1 Organizations and Businesses

One of the first things to do in any transformation is to find out what an organization and its businesses do. Existing business often looks towards expanding. Big local companies have been acquiring businesses across borders to become larger. Start-up companies see themselves as businesses that must generate funds from their ideas, as do individuals who must provide services to generate funds to run their households.

If you are involved in a transformation you might ask questions like:

- What are the businesses in the organization?
- What kind of activities does your business carry out?
- What other business do you interact with?
- What is your main product or service?
- What is the source of business funds?
- Who uses your product and service?
- Where do you want to be in the future?

After these questions, you might even draw a sketch, like that shown in ◘ Fig. 1.1, of your business and how it fits into organizations.

### 1.1.2 What Is the Role of Information Technology?

Most people are aware that technology has had a large impact on how businesses work. There has been a close relationship between information technology and ways of working in practice for many years. It began with using excel sheets to simplify accounting and today information technology impacts all activities in practice. The way it is used depends on the actual work practice, but also in developments in society and environment.

Technology is now increasingly used to provide the information needed by people in organizations to make decisions and manage their processes. Such organizations, which include individual businesses, cities, and communities, are now increasingly complex and are required to adapt to continuous change that introduces new challenges. Technology can be seen from two perspectives: improving communication between entities and improving the internal working of a business.

Increasingly, information technology is needed to help people and business and community networks to respond and adapt to change by dealing with emerging threats and opportunities.

## 1.2 Why Is Change Happening?

While finding out what an organization does, you might begin to find out what is disrupting its activities? Why is it changing? Again, initially you gather potential causes of a disruption from people in the organization, reports, and so on. You might find out what is causing the disruptions.

The term "disruption" is now increasingly used to describe changes in the environment that can affect our organization. The emerging terms are for networks to be resilient and agile in responding to disruption.

Any disruption will affect both the way each business works as well as how the relationship between them is affected by the disruption. ▶ Chapter 9 describes disruptions in more detail.

At the time of writing, society is going through a global pandemic known as COVID-19 that is impacting the way businesses are run and relationships between individuals. New ways of working are now evolving, while at the same time what has been learned in earlier phases has to be used to transform work practices to generate value for society. Increasingly, technology and science are playing an increased role in achieving transformations that add to social value.

### 1.2.1 Change Caused by Natural Disasters

The three drivers of change and thus disruptions are advances in technology, changing social values, and disruption—commercial and natural.

Natural disasters such as floods and forest fires are a common event in some areas, and organizations have been created to prepare for them. There are also preventative measures taken to reduce the damage that they cause. Taking such preventative measures is now common in many areas and is often referred to as disaster mitigation. It includes action plans that depend on the location and severity of the event, training of volunteers, and the affected community. Disruptions are covered in more detail in ▶ Chap. 9. ▶ Chapter 10 then describes potential disruptions resulting from climate change.

### 1.2.2 Change by Advances into Technology

Technology has now been the driver of change to both businesses and society. Such change is often driven by advances in technology, which is now an integral component of any system.

Uber and Airbnb are recent examples of change made possible by technology. Uber facilitates mobility in cities while Airbnb supports finding accommodation.

There are many other industries effected by technology. An obvious one is the retail industry, where businesses such as Amazon have had a profound effect on the way people shop. All these changes have called for the design of new processes that provide services that use technology, which provides easy ways to access services in simpler and cheaper ways. Cheaper and simpler ways to access services provides value to people—by reducing product costs, or reducing time spent waiting for a service.

Globalization is another trend where even the smallest businesses are seeking to becoming global to find markets for their products. Again, technology plays a significant role. Cloud technologies make it easier to store data in ways that is accessible anywhere in the world. Communication makes it easier to share such data between organizations. Social systems such as Facebook have extended communication to individuals creating new businesses, such as sole traders becoming social influencers.

### 1.2.3 Change Arising from Social Needs

There are now changes happening in society where population growth and lack of resources requires increasing support for people especially in large cities. Such support includes provision of services such as housing, or health for the disadvantaged to raise their standards of living.

Now you might ask questions like:

- Is the organization or business changing the way it works?
- What is causing the change?
- How does such change effect our main product or service?
- Who uses our product and service?

## 1.3 How Is Distribution Affecting People?

Ultimately, any disruption affects people. Increasingly, it is not only the commercial aspects that are affected but also what is called their values, and therefore their needs.

▸ Chapters 3 and 4 focus on values and what people think. ▸ Chapter 3 defines values and often suggestions and ideas commonly held by people. As you gather the information, you begin to gradually develop knowledge about the organization, how its businesses work, what values its people have and how they see their business and themselves affected. The amount of data collected here and its variety can be quite large. ▸ Chapter 5 provides a framework to organize the collected data into ways that can be used create the knowledge needed to decide where value can be gained from the transformation. It not only transforms the organization to deal with the current disruption, but also transforms it into a form that can deal with future disruptions. Then ▸ Chap. 6 describes how to combine what you have discovered about the people and organizations into a model. This model can be used to identify issues to be addressed, which is covered in ▸ Chap. 7.

## 1.4 What Is Needed from the Transformation?

Transformation requires designers to develop knowledge about an organization. It requires analysis of information to produce new insights that help decision-makers to focus on problems when addressing significant challenges. Now analysis goes into greater depth as, for example, matching people's demographics to their value preferences, thus identifying opportunities to identify services using such demographics. ▶ Chapters 3 and 4 describe how to gather information about values of businesses and their stakeholders, and ▶ Chap. 5 then describes how to use this knowledge in a transformation. Ways to develop models to describe this knowledge about an organization are then described in ▶ Chap. 6.

▶ Chapter 7 uses the model defined in ▶ Chap. 6 to precisely define the issues that any transformation must resolve and problems to be solved to resolve these issues. Before starting the transformation, however, it is important to agree on what issues it should address. Identifying issues and problems to be addressed now becomes important. Transformations are more likely to be accepted if they have a positive impact on people's and organizational values.

▶ Chapter 7 provides a way to identify problems by using people's values to evaluate improvements using agreed-upon best practices. Stakeholders evaluate the organization by comparing how adopting best practices can add value, be it commercial or personal.

Readers, of course, may want to start with ▶ Chap. 7 if they already have a model of their network.

## 1.5 How to Design Transformations?

▶ Chapter 8 then follows by describing how to design the transformation—how to change the system. Readers may wish to begin by reading ▶ Chap. 8 if they have already identified a problem. It describes design processes to create innovative solutions and ways to implement processes and technology to deliver value to stakeholders.

### 1.5.1 What New Business Practices Are Needed?

Increasingly, organizations no longer focus on well-defined problems. For example, should falling sales be seen as a problem or as a challenge to find and retain customers. The question then is what shall we do, what problems to address the challenge? For example, the problem with reduced customer growth may be due to poor product performance, or increased competition, or delivery issues, or maintenance support. These present challenges rather than problems to organizations. How to address new competition is often seen as several challenges—improve product, improve marketing, and improve access to the business.

Businesses must now address wider issues. Increasingly, they must address issues that arise from disruption, especially resilience and sustainability. To become resilient, businesses must develop ways to identify the problems that emerge from a disruption and solve them in ways that deliver value both personal and commercial. For example, reducing traffic congestion in cities by addressing personal values such as reducing stress by freeing up time.

### 1.5.2 Continuing with Globalization

Global trade now impacts everybody's life (Rowntree et al., 2017), with growing emphasis on global supply chains that can deliver goods to individuals in increasingly faster times. Such improvements in supply chains are the result of digitization, where technology is used to deliver services in increasingly ubiquitous ways anywhere in the globe. There is now increasing pressure to continually improve services. Innovation, creativity, and entrepreneurship are now playing a greater role in all activities. Innovation itself is leading to new ways of working where, through collaboration, we can put knowledge together to create new services.

### 1.5.3 Fostering Innovation and Creativity

The emphasis on innovation and creativity that has taken place is now increasingly important. Such innovation has been illustrated in responding to the global COVID-19 pandemic as, for example, the use of technology in widespread remote work. There is even greater emphasis on developing new ways of working to maintain the living standards that people have been accustomed to. One obvious example is working from home. The question here is how to maintain productivity while doing so.

### 1.5.4 Changing Work Practices

Disruptions call for new ways of working. Perhaps one of the major effects of the COVID-19 pandemic has been the growth of remote work; and the resulting technological developments and social practices. There is no doubt that many of these practices will remain after society learns how to live with the pandemic. How will work be organized and how will technology support any new work practices? Will the need for business travel decrease? Will new virtual communities or business arrangements be formed where the participants are those that are globally, not locally, based?

### 1.5.5 Changing Services and Products

Increasingly, the term *resilient or sustainable services* is becoming relevant. Any change in work practice calls for new work practices and delivery of services. The book places considerable attention to developing the journey, where every point in the journey can be carried out as independently as possible from others. Such independence leads to resilience, where any part of the process can be easily moved to a more favourable place. Journey maps are introduced in ▶ Chap. 6 to model processes as part of the business model.

### 1.5.6 Making Business and Cities Smarter

The term "smart" is now increasingly used to describe ways that technology, and especially data, can lead to better business outcomes as well as developing better living environments. The term smart is hard to define, but basically the word smart stands for actions that are *S*pecific, *M*easurable, *A*chievable, *R*ealistic, and *T*imely in creating new services or products that create value. Data is considered as a crucial resource here, and technology is increasingly used to align data to the decisions with people needs. Increasingly, artificial intelligence is being used to identify ways to collect and use data to make smart decisions.

### 1.5.7 Supporting Knowledge Creation and Innovation

Now information technology not only provides information but is increasingly used to create new knowledge to present insights that can lead to new ways of work. The growth of knowledge has continued over time. There is now increasing volumes of information becoming widely available through Google, Facebook, and other sites. Information is now widely available and other generally available providers. It is, however, still needed for people to make sense of large volumes of data in their everyday work. Questions such as how can I use new knowledge to improve the way we work? How do I organize information to create the knowledge needed for better decision making?

### 1.5.8 Capturing and Analysing Data

Data is needed to enable decisions that make systems smarter. Such data is becoming increasingly available through technologies, sometimes known as big data, and through the Internet of Things (IoT). Big data is increasingly seen as of great value in identifying business trends and people preferences. The goal then is to develop the intelligent systems that can be used to transform business systems to create value in effective ways. The internet of Things, on the other hand, are technologies that are increasingly used to monitor what is happening in cities and create the data needed to make decisions such as improving traffic flow or increasing safety.

Data captured by the Internet of Things (IoT) must be summarized and presented in forms suitable for decision-makers. Artificial intelligence is also increasingly embedded in systems to assist decision-makers and the community.

### 1.5.9 Developing Living Environments

Technology is also becoming a critical factor in the design of ever-growing cities. Examples include transportation schedules like bus timetables or online shopping. It is here that social issues are beginning to take increased importance and that data can make designers aware of what the issues are in both businesses and cities. The focus is increasingly on providing ways to build systems that address people's values.

## 1.6 What Is Important in Transformation?

The term "value" is now emerging more and more in organizations. This book will describe value in more detail in ► Chaps. 3 and 4, but basically it is something that people need and like. Any disruption must create value to most people, Airbnb delivers value because it gives people access to more and more reasonably priced accommodation in a greater range of areas. Online shopping provides value by saving busy people time and, in many cases, paying less. This book focuses on transforming systems to deliver value.

Transformation is seen as a process whose goal is to deliver such value through innovation. Transformation includes business systems, cities' personal needs, health, and support for people in cities. Increasingly, technology provides the connection between people. There is also general agreement that such connections are not only useful in growing a customer base, but also can lead to innovative outcomes. As a result, here is an increasing trend to digitization and globalization made possible by technology, where technology is leading to greater sharing of knowledge.

### 1.6.1 Social Issues

Sharing economies are often seen as themselves causing disruptions—they affect current operations and livelihood of others—existing taxi drivers, in the case of Uber (Boshuijzen – van Burken & Hafor, 2017). What are the disruptions? In the past, it was most often a new competitor. Now it is the whole industry changing. For example, Uber has had several impacts often seen as disruptions to the taxi industry, including the following:

- Taking jobs from existing taxi drivers.
- Not being regulated to provide safety.
- Uber drivers do not need to be licensed to provide taxi services, whereas taxi drivers do at their own cost.
- Uber can track customers, which does not happen in the existing taxi industry.
- Many of these give Uber an advantage over existing taxi drivers.

Social issues go beyond Uber, but are now typical of what is now commonly known as the "gig" economy. Many businesses here rely on having a pool of workers on whom they can call virtually at any time. Some of these have resources to provide for the business, like car owners in Uber, while others, like food delivery drivers, only have their time to sell. Exploitation of food delivery drivers is now an increasing issue in society.

The focus now is not on just using technology to improve business processes but, at the same time, use it to address social issues, both at the personal and organizational levels. Design covers not only designing business systems but also increasingly to add value to people's experience while working in the business. The goal is to align people values to business values. Matching values not only applies to customers—it also applies to employees. Successful businesses now build systems that combine the two—both satisfy people and city values.

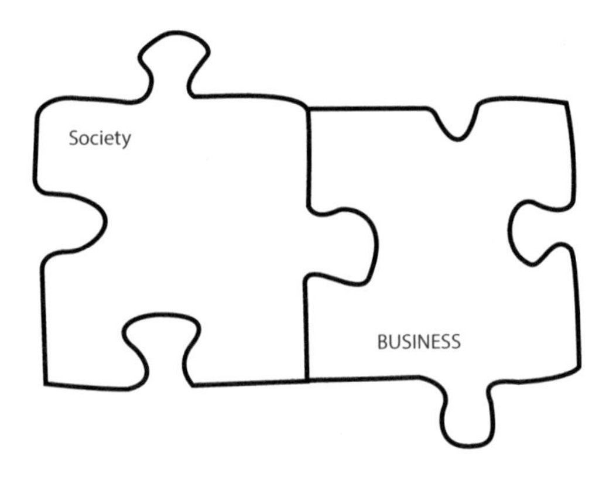


Building systems that include information processing is also getting more challenging. What is increasingly becoming important is that such systems must be designed to manage continuous change, both social and business, and maintain a good balance—both in its internal operations but also in responding to changes in its environment.

## 1.7 The Role of Technology

Technology, especially information technology, is now seen as an integral part of any system. Whenever you buy a ticket to travel, or make a purchase at a supermarket, technology is part of the process. It not just simply issuing the ticket or collecting payment; it now also analyses travel patterns or purchase preferences to help business decision-makers. Information technology is of course continuing to play a major role in business. It is now used extensively in any firm's everyday activities. It supports business processes by quickly moving information and making it available at the right place to the right people, which include both customers and business employees. Increasingly, it is used to assist decision-making using data analytics tool that show emerging relationships in the data.

### 1.7.1 Improving Efficiency

Earlier, information systems' design focused primarily on business processes. The focus was on what is called structured data often focused on one task, such as placing an order. Another was making paper records and placing them on Excel sheets that created efficiencies that led to much relief from routine activities so that now people can focus on more innovative activities. Now systems can track supply chains to maintain awareness of the delivery status and provide information on the status of any projects.

Technology is a two-faced tool. It can be used to improve efficiency, but also it can help solve problems and lead to new ways to conduct business, and act as a disruptor to existing businesses. It is not only automating what were previously manual processes but providing new ways to conduct businesses. Airbnb and Uber have been successful because of opportunities provided by technology platforms where people make their own arrangements to provide services. These platforms bring people together to share resources or provide services. Such platforms have been available over several years in systems that bring people together to provide services for each other, as for example, dating services.

### 1.7.2 Delivering Services Using Technology Platforms

Information technology is increasingly used to create platforms that provide the services needed by people. Services not only satisfy each person's needs but must coordinate their activities to avoid non-productive time in waiting or repetitive work. Most coordination here is often still using e-mail. New tools such as dashboards are increasingly used to allow people to keep track of what is going on and take the necessary action when needed.

Platforms should be ubiquitous and natural. Just like phones are now. In the early days, to make an international call required considerable knowledge about the process of connecting to the caller. You needed to know that you call an operator and request a connection. You needed to tell the operator details of the caller. If so, you were connected. The operator would then contact the caller to see if they are willing to accept the call. Now you pick up your mobile and dial the number. You do not need to know the process followed. The platform should know the process steps and alert people of actions they should take.

Thus, in ◘ Fig. 1.2, a typical system now includes a set of application programs that support business activities. The platform provides services, which support business systems through computer interfaces to both service providers and recipients. The way the data is stored, and internal operations are not of any concern to users—users just need some service. There are many examples of platforms now in industry.

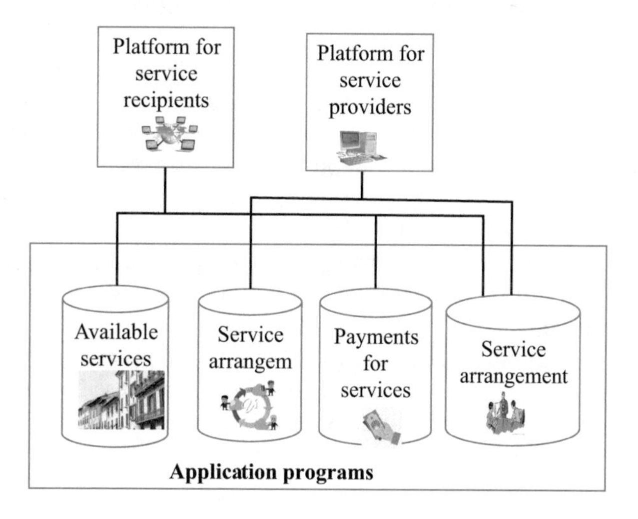


◘ **Fig. 1.2** Delivering services

These include platforms for on-line purchases, registering for government services, making reservations, among others. Facebook can also be a platform, providing services for maintaining relationships.

The role of technology is thus seen here as providing invisible platforms through which people can access services that deliver value. The platforms are supported processes that use an organization's resources to deliver such services.

These are focused on supporting collaboration, but as people realize the growth of remote work, more elaborate platforms that integrate collaborative activities into corporate systems will no doubt emerge.

### 1.7.3 Technology Trends

Technology is one important factor contributing to business model change. For example, facial recognition can lead to creating value through avoiding queuing—very few people like to stand in queues. Examples of providing value through facial recognition, include the following:

- Providing a seamless process as, for example, going through several check points while entering a flight, made possible by facial recognition.
- Shopping without checkouts with all items in the shopping trolley scanned as you exit the supermarket.
- The role of drones in supply chains or local deliveries.
- Increasing use of artificial intelligence.
- Robotics becoming more common.

At the same time, however, technology can contribute negatively to privacy, which is an important value to people. Designers must somehow design business models that maintain privacy of people's facial descriptions, while delivering value through reduced queuing time.

## 1.8 What Are the Emerging Organizations?

Organizations now face an increasingly changing environment and must be resilient to unexpected change and disruption. ▶ Chapter 9 focuses on disruption in greater detail, noting that businesses must include ways to mitigate against disruption and recover. ▶ Chapter 10 describes potential disruption from climate change, and ▶ Chapter 11 how businesses operated in ever-growing cities and how cities themselves deal with disruption while businesses operate in the city. New important challenges are emerging for cities—smart cities and what is increasingly seen as important—climate change. What kinds of applications or business models are needed here?

Business models must continually adapt to changes in the business environment. The changes in the way are leading to increased complexity. The business examples described so far may only be the tip of the iceberg. The final ▶ Chap. 12 describes trends in business models.

### 1.8.1 The Sharing Economy

The sharing economy (Acquier et al., 2019) is one example of new ways of conducting business, or as emerging new ways of doing business. It addresses the opportunity or challenge of utilizing idle resources. Systems such as Airbnb or Uber now provide platforms that put together people where one person provides a service to another person managed by the platform—such ways of working are now commonly known as the sharing economy. They emphasize ways to use otherwise unused or idle resources to create new business, such as cars idle in the garage or empty rooms in case of Uber and Airbnb. The two best-known examples are Uber, where private cars normally in garages are now used to transport people, or Airbnb, where spare rooms can be used to house people.

The sharing economy is also one that is seen as causing disruptions to existing systems. Such disruptions now often occur through the combination of digitization and globalization. Airbnb is a typical example where a technology has made it possible to globalize a service where people in one location can provide accommodation to others anywhere in the world. These are often referred to as complex systems. ► Chapter 2 describes what is known as a complex environment and then describe the design processes needed in such complex environments.

### 1.8.2 The Gig Economy

The term "gig economy" is also emerging. Here, individuals, who are sole traders, offer their services across several organizations. The term "digital nomad" is also now more often used to designate people like contractors who take transient but often highly skilled jobs. The trend to remote work is likely to increase as now specialists with skills will be able to travel "virtually" rather than physically, especially in areas such as planning, financial advice, and in design.

The gig economy can go beyond sole traders delivering meals, but increasingly incudes sole traders or small teams organizing themselves to deliver high level services.

### 1.8.3 Smart Cities

It is not only businesses that have to work within complex environments, especially cities. The growth of cities contributes to the complexity as most businesses and organizations must provide services in city environments and contribute to city values. This book describes city values and how businesses and organizations contribute to city values. These are described in more detail in ► Chap. 11.

### 1.8.4 Dynamic Business Networks

Another emerging trend has been the formation of business networks. Business networks and alliances have existed over many years, the dynamic nature of today's disruptive environment has meant that such alliances can emerge quickly and be of shorter duration. Furr et al. (2016) explains the dynamic nature. Furr and others (2016), for example, describe ways major companies can work together on developing new capabilities across industries.


## Summary

This chapter described general trends in business and society and their impact on information system design. It identified value as an increasingly driving factor in design and the increasing need to design systems that can deliver value in an increasingly dynamic world.

This chapter also indicated two emerging issues in transformation—complexity and value. The next chapter, ▶ Chap. 2, will provide an overview of dynamic complex information systems, how to describe complexity and challenges to be addressed by designers to create designs that manage complexity in the most effective way. It describes the importance of focusing on stakeholder values, identifying what we can do to realize values, and the variations and changes that any systems must address.


## Exercises and Case Studies

It is most important to find out how an organization or business works before proposing any changes. You can only find out how organizations work by asking questions and recording answers.

Following are three case studies. Ask questions like those found in this chapter? What are the businesses? Where do they acquire resources? Which businesses use their products and services? As you are collecting answers, draw a diagram like that shown in ◘ Fig. 1.1 to show the different kinds of businesses, people that are part of the organization's environment. Do not restrict your questions to the organization but also look at their environment—for example, what similar organizations are doing, or what are the trends in the environment.

### Shopping Centre

Most people are familiar with shopping centres. Many see shopping centres as an arrangement that provides convenient services to shoppers in one location. They also see the shopping centres as an organization that includes many shops, which are businesses that make up the shopping centre. These businesses see the shopping centre as an organization and must ensure they remain competitive in their environment.

### Restaurant

What is a similar environment diagram for restaurants? Look at the relationships between restaurants and their customers, suppliers?

What kind of disruptions will affect them? Include disruptions of all kinds, such as commercial challenges by competitors, changes to eating habits, and including those created by pandemics where the focus is on health.

What kind of services do they provide? How will they make their services resilient to changing tastes and trends? And what about the safety concerns emerging with COVID-19?

**Food Supply Chain**

Here it may be necessary to look at different businesses in the food supply chain—farmers, processors, supermarket, and restaurants are examples. Pay special attention to the coordination between the different businesses.

For each of the above:

- Ask the questions found in this chapter to familiarize yourself with the organization.
- Record stories either as post-it notes or a table like that below.
- Draw sketches or diagrams like Fig. 1.1 showing how their businesses are organized and what they provide for their customers or interact with other businesses.

Table for stories

| Story/ article | What is interesting about this story | Where did the story come from |
| --- | --- | --- |
| | | |
| | | |
| | | |

## References

Acquier, A., Carbone, V., & Masse, D. (2019). How to create value(s) in the sharing economy: Business models, scalability, and sustainability. *Technology Innovation Management Review, 9*(2), 5–24.

Boshuijzen – van Burken, C., & Hafor, D. M. (2017). Complexities and dilemmas in the sharing economy: The Uber case. In *Designing and managing* (pp. 1–11). Linnaeus University. https://doi.org/10.15626/dirc.2015.04

Furr, N., O'Keefe, K., & Dyer, J. H. (2016, November). Managing multiparty innovation. *Harvard Business Review*, pp. 76–84.

Rowntree, L., Lewis, M., Price, M., & Wyckoff, W. (Eds.). (2017). *Globalization and diversity: Geography of a changing world*. Pearson Education.

# Transforming in Complex Environments

## Contents




I. T. Hawryszkiewycz, *Transforming Organizations in Disruptive Environments*,
https://doi.org/10.1007/978-981-16-1453-8_2

This chapter sees transformation as making choices in increasingly complex environments. It describes what complexity is and how it effects transformation. It shows why transformation must proceed in an incremental manner learning about complexity as transformation proceeds. It also describes the importance of developing knowledge about the organization as transformation proceeds.



The chapter concludes by describing how to organize a collaborative team to develop such knowledge and contribute to the transformation. What you will learn here is how to arrange your team to carry out a transformation in small steps.

**Learning Objectives**

- What is complexity?
- What are wicked problems?
- Importance of collaboration
- Cities as complex systems

## 2.1 Introduction

If you were to go back just 60 years, life seemed simple. For shopping you went to local shops and bought what you need. You went to work often by catching a bus, arriving at your place of employment at 9, and then left at 5. The choices you made in your work were relatively simple and often based on simple rules or applying well-defined skills. Life has become more complex, and choices are now not so simple.

### 2.1.1 What Is Complexity?

People talk a lot about complexity—what is it? Complexity is often used as a general term, without explaining why a particular system is complex. Complex systems are seen here as made up of components and relationships (or connections) between these components. These may be business systems. There is often a variety of both connections and components. These relationships can be static or dynamic. The difference is that in static relationships, irrespective of the number of connections, it becomes possible to predict the behaviour of a system. Accurate prediction is often the case in engineering systems. In social systems, accurate prediction is not often possible as people's values and needs continually change. Such change leads to uncertainty in system design and calls for providing a greater variety of responses to deal with the variety as well as disruptions that often take place.

Another common term here is complicated. Complicated systems are different from complex systems. Complicated systems are those where there are many parts, but the behaviour of each part is predictable. There are many connections in complicated systems, but they do not change, they are static. It is possible to determine the best connections for given goal—for example, combining parts to create a car engine. Dynamic complex systems are different—their behaviour is not predictable hiring a new person can result in unanticipated behaviour by other people and how they see and use systems.

Definition of Complex Information systems

- There are many intricately connected *businesses and people* interacting with each other. These include farms, resource suppliers, and transportation systems.
- The businesses continually interact.
- There is a large volume of information in the environment.
- There are many stakeholders with continuously changing relationships.
- Changes are continuous, often frequent, and unexpected, often caused by disruption.

The key factor that makes social systems difficult to design is that they are *dynamic* not *static*. In many engineering systems, relationships do not change—the relationship between a wheel in a car and the car engine is predictable. Such prediction is not possible in social systems—replacing one person by another can impact on outcomes because of conflicts between stakeholders. For static systems, ways and models exist to represent static complex relationships and predict their behaviour. In dynamic environments, change introduces uncertainty, and systems must be able to manage uncertainty. Such systems are often called agile as they can respond to change as, for example, *changing* customer needs. Examples include the following:

- Greater agility in business and industry—responding to changing stakeholder needs and disruption.
- Raising health outcomes—täking into account doctor expertise, hospital management, patient management, prevention, and identification.
- Smart liveable cities—energy, traffic, water, education, entertainment, food, accommodation.
- Crisis management—facing natural disasters.
- Online learning—assessment, delivery, content.

Design is focused on meeting the values of many stakeholders. Such values can themselves change, sometimes very quickly, and business need to increasingly organize to identify and respond to continuous change.

## 2.2 Wicked Problems

Wicked problems are a term increasingly used to design in complex environments (Buchanan, 1992). Wicked problems were initially defined by Rittel and Webber (1973) in the context of planning (Camillus, 2008) and have increasingly been found to be applicable in describing problems in complex environments. Wicked problems (Ritchie, 2011) are characterized by the following:

- There is no definite specific formulation of the problem; there are just general goals such as increased sales in a new market, everybody well fed, or increasing tourism in some region. Often different stakeholders may have different versions of what the problem is. It is not possible to define phases or stages of design.

- There is no stopping rule—for example, when can we stop research that leads to better health. Often design stops when a deadline is reached, or people run out of time or money.
- Solutions are not true or false, but better or worse. They are judged by stakeholders, who, for example, may find them "better than now" or "good" or sometimes "unsatisfactory." Any measures are usually qualitative rather than quantitative.
- There is no test of whether a solution will work. It is not like chess, where you can try something and go back. You can only test if a solution will work if you try part of it. But in most complex situations, you cannot then go back. Often solutions lead to changes in behaviour, which requires further change.
- Every solution is unique, and solutions that apply in one environment cannot be used in others.
- There are many possible solutions.
- Every wicked problem is unique—every city needs a different solution to become smart.

A common way to view a wicked problem is like undoing a set of tangled strings as shown in ◘ Fig. 2.1. The goal is to untangle them into a set of separate strings and join them into a more organized way. Each string is some activity, and the tangle needs to be assembled in ways that activities can be put together to work in a better way, maybe connected into several long-connected strings.

Solutions to wicked problems as follows:

- Are not simple.
- Are developed gradually as we learn and must continually improve.
- Cannot be provided by technology alone.
- Require changes in stakeholder behaviour.
- Require adapting continuous changes to solutions.

◘ **Fig. 2.1** Wicked problem as a set of tangled strings

Furthermore, it is often not possible to create the best solution as was the case with structured deterministic systems. There is now a greater emphasis on creating value, which requires providing continuous support for change.

### 2.2.1 Why Is Complexity Growing?

Complexity has always been there (Merali, 2006). The difference is that now change is happening much more quickly, often through the introduction of technology. Complexity is growing because of continuous change in environment and increasing disruption. This change is caused by factors such as population growth, movement of people to cities, and the growing importance of education, and recently, COVID-19, all of which require new resources and services. Examples include the following:

- Changing relationships in business systems to deliver services to cities, towns, and villages.
- Disruptions that require change in organizations and behaviour.
- The knowledge in each business system continually grows.
- Leading edge technologies such as cloud, mobility, big data, or social software leading to new ways of working and competition.

But perhaps what is most important is that people's values are continually changing. The continually changing environment places new challenges in design. It is no longer possible to specify what are commonly known as user requirements and create a system that is deterministic in nature. Any new system must now be able to deal with environments with continuous change. Such ability must be built into the design. The kind of abilities generally found as needed in changing environments include the following:

- The most well-known one is that of *requisite variety*. Basically, requisite variety says that a business must have a way to deal with any expected change. Thus, do not design a business to deliver a fixed service to a known group of clients—design it so that it can deal with changes in client needs.
- *Feedback*—information about the effects of any previous actions.
- Creating the ability to *self-organize* in response to a change or to a disaster situation.
- *Emergence*—new discoveries are made as complexity is understood. These result in new system activities.
- Develop ways to *frame* a problem by creating a boundary around that part of the system affected by proposed changes.
- Be clear on the *purpose* of any change.

The ability to manage change goes beyond simple businesses. How then does working with wicked problems impact the way we work and ultimately transform our work and behaviour.

Consequently, it becomes necessary to transform our activities to work within wicked environments and transform to systems that address the challenges of wicked environments. Questions to ask here include the following:

- Do you and your systems have different ways to respond to unexpected events?
- Do you use feedback from your connections?
- Do you have ways to look for new opportunities?
- Can you quickly rearrange the way you do things?

## 2.3 Impact of Wicked Environments on Transformation

Communities and business networks have been continually transforming the way they work in wicked environments. The difference recently is the nature and speed of transformation. In fact, some people see that we are continually transforming by designing new ways to do whatever we do. Such transformations take place at individual and community levels. Some people, for example, see smart cities as communities that are continually redesigning the space they live in.

The greater emphasis on social values is one characteristic in wicked environments, where it is often not possible to predict exactly what will happen when we put things together, especially in social environments, where changing people's responsibilities or adding a new person can often lead to unanticipated outcomes.

### 2.3.1 Building in Small Steps and Learning

Another characteristic of wicked environments is the formulation of the problem is itself often the problem. It is easy to say that the problem is to develop a strong customer base. However, because of the complex relationships it is not possible to develop ways to solve problems in a deductive way. Instead, we go in small steps like that shown in ◘ Fig. 2.2. It is like unravelling a string one knot at a time, not knowing what new knots you will discover. Rules have limited value here and make deductive reasoning (Redante et al., 2019) difficult. In ◘ Fig. 2.2, for example, the goal may be to improve health in a city. There is no one specific outcome, but we proceed in small steps, just like untangling strings, continually improving different health

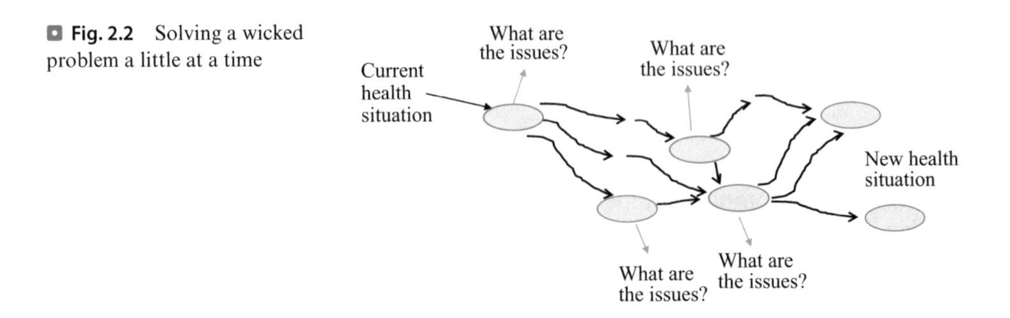


◘ Fig. 2.2 Solving a wicked problem a little at a time

issues towards a better city health outcome. At each step, we identify the most urgent issues and find new problems. Once they are addressed, we go to the next set of newly emerging issues. In a pandemic, like COVID-19, for example, tracing is used to find hotspots and respond to them—framing the problem to the hotspot. The problem of improving health never stops. The steps depend often on opportunities that arise, identified specific health issues, and funding availability

### 2.3.2 Being Creative and Innovative

Solutions in wicked environments call for new ways of carrying out activities by developing new experiences for your business or its customers by being creative and innovative. Some writers see the difference as between creativity and innovation. Creativity is seen as doing something not possible before. Innovation is often seen as finding better ways to make it possible. Solutions often require both creativity and innovation. One could argue, for example, that Airbnb is both—providing home-owners with new experience in renting their property, while being innovative by providing a platform through which they innovate using technology.

## 2.4 Collaborative Team Activities

Collaboration is another practice that is encouraged to foster creativity. We need to share any knowledge identifying the emergence of new situations. The emphasis here is often on finding a response and on multi-disciplinary teams to use their joint expertise to find that response. Collaboration is not only people continuing discussion (Perlow, 2017). Collaboration is an organized way to collect the knowledge needed in design, make sense of it by identifying issues that need improvement, and then create innovative solutions. Designers work in teams if practicable, and collaborate to find out and discuss how businesses work now and their ability to respond to change. Collaboration involves people with the knowledge about the organization as well as people with knowledge about ways to address and solve problems. The nature of collaboration changes (Bernstein et al., 2019; Cross et al., 2019) as transformation proceeds. As shown in ◘ Fig. 2.3:

- Learning about the organization collaboration, which calls for analytical skills, and knowledge in the subject area.
- Finding out issues or problems that need to be addressed, calling for analytical skills combined with creative skills in identifying possibilities.
- Creating solutions that solve these problems that calls for creative and innovative skills.
- Implementing and evaluating solution.

Often collaborative teams change as transformation proceeds. Initially the teams are made up mainly of people internal to a business as they know most about the current

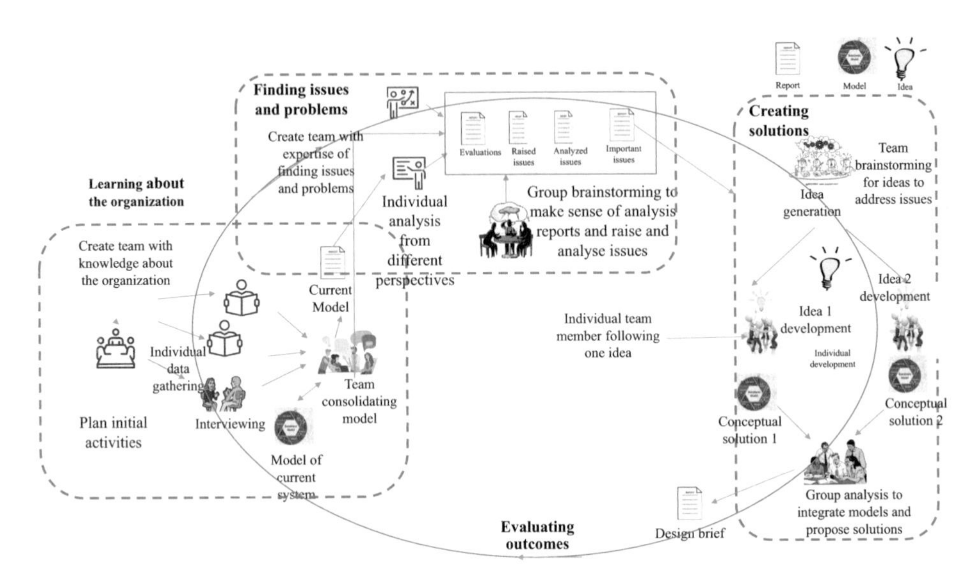


Fig. 2.3 Organizing collaboration

working of the business. Then, as collaboration proceeds, new people can be brought in with ideas of how things can be improved. Such change is often through good leadership.

## 2.4.1 Leadership

Transformation requires good leadership. Good leadership organizes teams in ways they can be managed to develop creative and innovative outcomes. Research in creativity (Amabile & Pratt, 2016) has identified four components shown in Fig. 2.4, whose combination leads to creative outcomes. These are as follows:

- A social environment needed to support creative activities that spontaneously lead to new ideas, while bringing in knowledge from different disciplines.
- Creativity relevant processes that look at a problem from different perspectives, while encouraging inputs from a variety of sources.
- Motivation—intrinsic and extrinsic.
- Expertise or domain-relevant skills to ask critical questions.

Leadership of networks must organize the business teams by the following:

- Employing people with the analytical capabilities focusing on current business problems.
- Organizing resources—technology, data analysis processes, high quality data, a data warehouse that makes it easy to analyse data relationships.
- Developing ways to motivate through reward systems and employee empowerment.

Collaboration goes beyond businesses, but is important when creating liveable environments through community activities.

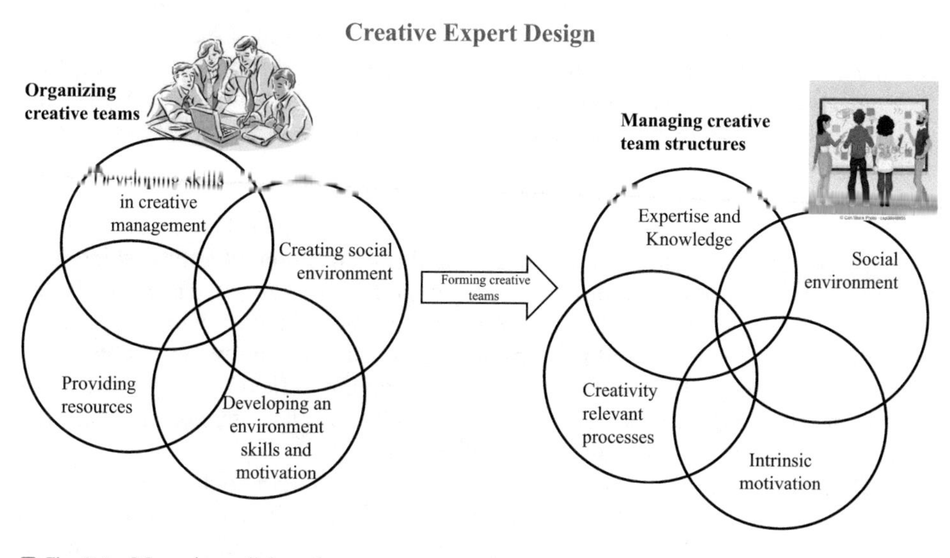


Fig. 2.4 Managing collaboration

## 2.5 The Importance of Communities

The pandemic of 2020 has been amazingly effective in defining the concept of communities and the need for communities to act in unison to address the problem of the transmission of the virus. If anything, experiences like those in Italy or Spain have drawn attention to its damaging effect. In the face of the pandemic, jurisdictions have increasingly sought community consensus for ways to deal with the pandemic. Such consensus has happened at all levels. It has brought forward the importance of leadership that develops creative communities with shared goals.

Collaboration itself requires communities that reach consensus on a common goal. The need for such consensus is illustrated by cities that are large communities. Such consensus has been needed during the COVID-19 pandemic, where disease, which is carried by humans, finds it easy to spread through people in confined spaces such as large building, or entertainment values. The response in many cases requires lockdowns, which requires consensus that many find burdensome, as movement has been a clear social and business value.

### 2.5.1 Cities as Communities

Most people are now aware that world population is growing, while at the same time there is a movement of people from rural areas to the cities. The term megacities (Chen et al., 2018) is now often used to describe the rapidly growing cities. There are now about 50 cities in the world with a population of over 5 million and 10 with a population of over 12 million, with the largest being Shanghai (24 million). Australian cities are small compared to some megacities (Sorensen & Okata, 2011) with Sydney, the largest, with a population of 5,000,000 and classed as the 56th largest in the world.

The growth of population and its movement to cities put pressure on cities to continually improve services to citizens, but also to respond effectively to disruption through improvements in health, education, and mobility, among others. At the same time, there is increasing concern of climate change impacts that is increasingly calling for such services to be provided in a sustainable way. Increasingly, such services can only be provided through smarter use of technology.

Prior to the pandemic, the idea of smart cities has emerged. The goal of a smart city is to provide increasing services made possible by technology. Cities, which have the goal of being smart, need not necessarily be today's megacities. They can also be smaller cities or townships that want to improve liveability by improving city services and amenities in their city and perhaps attract industry and grow. In this book, we see cities as increasingly complex and design as solving wicked problems in wicked environments. The general approach here is to solve problems by decomposing the problem to manageable parts and solve each separately.

▶ Chapter 11 describes the idea behind smart cities in more detail. For example, simply improving hospital outcomes does not raise the health outcome if people find it difficult to get to the hospital. Thus, improving mobility becomes part of the design process. The problem boundary must include both the hospital capability as well as the transportation system to the hospital. The purpose becomes getting sick people to hospital as quickly as possible. The question then becomes how to design the best way to improve access to hospitals. These are discussed in ▶ Chap. 11.

### 2.5.2 Emerging City Values

There has also been interest in finding measures that characterize good city environments. Such measures have been continually emerging to define and compare smart cities. These are often specific measures that impact more on citizen behaviour, with emphasis on supporting ways to lead to well-being and the creation on innovative communities. Some measures are common across cities. These include generic issues such as mobility, health, safety, and productivity such as:

- Walkability for health, commerce to contact others and carry out commerce today increasingly maintaining social distance.
- Safety to ensure people can move in their communities without fear.
- Mobility to ensure that people can move within the community.
- Sharing resources through collaboration to support communities working together.

These measures become some of the best practice themes in ▶ Chap. 11 when smart cities. Chapter 11 also describes policy development for health, accommodation, or energy use which often call for communities themselves to deal with in innovative ways.

Ways of developing such policies will be described in more detail in ▶ Chap. 11. However, it may be worthwhile to point out here that of the generic issues above, only mobility conflicts with recommendations for responding to the COVID-19 pandemic. Mobility of people is seen as spreading the virus.

### Summary

In summary, designers can no longer prove that they have the best design using quantitative measures. They must increasingly rely on proving that their designs add value to people, and they can respond to changes in the complex environment.

This chapter described what is meant by complexity in the context of information systems and defined some guidelines for design. The next chapter begins to describe values of people.

There is also a mutual relationship between business and city where each should contribute to each other's value. Ä city provides services that enable businesses to flourish in the city. A health service öperates a business while contributing to the health of a city.

### Exercise

Continue with the case studies in ▶ Chap. 1. You might want to work in teams if practicable and collaborate in finding out and discussing how the businesses work now and their ability to respond to change. Use the guidelines in ◘ Fig. 2.4 when creating the team, making sure there is someone who has some knowledge on the case study.

For each of the case studies, look at each business and draw the relationships between it and other businesses.

- What information is exchanged in the relationship? How can the information change?
- What would you suggest the business do to anticipate and respond to the change?
- Does the system satisfy the definition of complex system?
- Focus especially on the changing nature of relationships.
- What kind of abilities would you provide to respond to change?

Add to the stories you have recorded in ▶ Chap. 1.

## References


Amabile, T. M., & Pratt, M. G. (2016). The dynamic componential model of creativity and innovation in organizations: Making progress, making meaning. *Research in Organizational Behaviour, 36*, 157–183.

Bernstein, E., Shore, J., & Lazer, D. (2019). Improving the rhythm of your collaboration. *MIT Sloan Management Review, Research Feature, 61*(1), 29–36.

Buchanan, R. (1992). Wicked problems in design thinking. *Design Issues, VIII*(2), 5–21.

Camillus, J. C. (2008, May). Strategy as a wicked problem. *Harvard Business Review*, pp. 99–106.

Chen, C., LeGates, R., Zhao, M., & Fang, C. (2018). The changing rural-urban divide in China's megacities. *Cities, 8*, 81–80.

Cross, R., Davenport, T. H., & Gray, P. (2019). Collaborate smarter, not harder. *MIT Sloan Management Review, Research Feature, 61*(1), 20–28.

European Union Regional Policy. *Cities of tomorrow*. citiesoftomorrow.final.pdf

Merali, Y. (2006). Complexity and information systems: The emergent domain. *Journal of Information Technology, 21*, 216–228.

Perlow, L. A., Hadley, C. N., & Eun, E. (2017, July–August). Stop the meeting madness. *Harvard Business Review*, pp. 62–70.

Redante, R. C., de Medeiros, J. F., Vidor, G., Cruz, C. M. L., & Rebeiro, D. (2019). Creative Approaches and green product development: Using design thinking to promote stakeholders' engagement. *Sustainable Production and Consumption, 19*, 247–256.

Ritchie, T. (2011). *Wicked problems – Social messes*. Springer.

Rittel, H. W., & Webber, M. M. (1973). Dilemmas in the theory of planning. *Policy Sciences, 4*, 155–169.

Sorensen, A., & Okata, J. (Eds.). (2011). *Megacities: Urban forms, governance, and sustainability*. Springer.

# Values to Be Achieved

Contents

# Importance of People Values in Transformation

## Contents




I. T. Hawryszkiewycz, *Transforming Organizations in Disruptive Environments*,
https://doi.org/10.1007/978-981-16-1453-8_3

Successful transformations change organizations to deliver products and services in better ways. They change existing services, develop new services, and provide better ways to use technology to deliver such services. The terms often used is that a good transformation uses technology to simplify delivery of services that deliver value to people. This chapter describes values that people may hold. It focuses on what people value as important and how to create new value.

Finding knowledge about people's values is where we ask "what and why questions" to gather necessary information. This chapter also introduces the first steps of gathering information about people's concerns about what is happening now and what they think should be done.

**Learning Objectives**

- What are values?
- The triple bottom line
- Aligning personal and organizational values
- Developing Persona maps
- What are stakeholders?

## 3.1 Introduction

▶ Chapter 2 noted that it is no longer possible to design transformations by using rules to make choices—that is, methods based on formal rules and outcomes that can be computed as optimal in some way. The emphasis is on delivering personal and organizational values, where rules are not explicit.

Values themselves can be studied as a discipline. It is not the intention here to go in depth in discussion on the theory of values, but to focus on their relevance to organizational transformation. Ultimately personal values must be aligned to organizational and business values, and increasingly to environmental values, leading to what is referred to as the triple bottom line. ◻ Figure 3.1 illustrates the idea of triple bottom line, where transformations must satisfy three sets of values—people, business, and the environment.

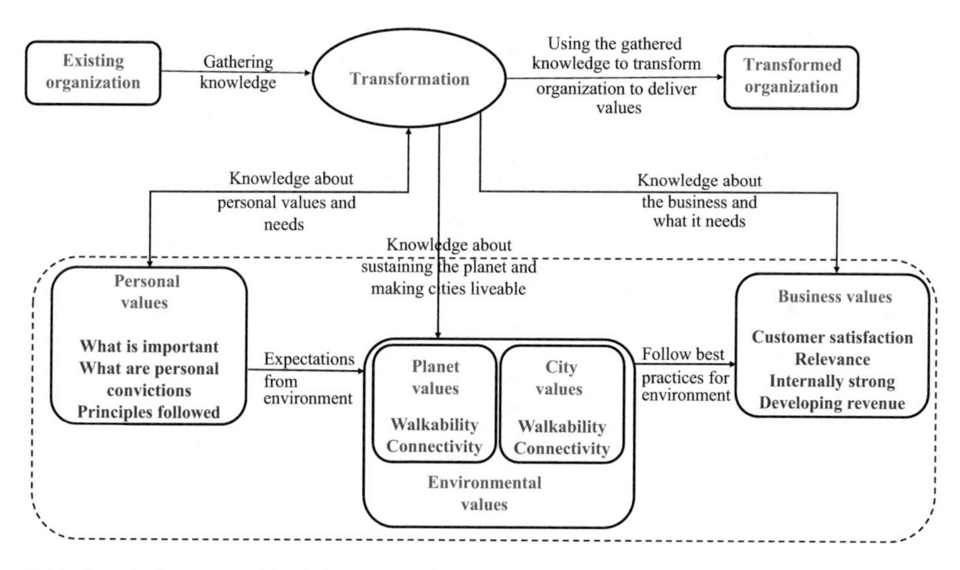


◻ **Fig. 3.1** Values to consider in transformation

The combination of these values is illustrated in ◘ Fig. 3.1, which shows the kind of knowledge gathered about people, businesses, and their environment. These include:

- Personal values of people in general, by identifying what they find important, what they are convinced is true, and what principles they follow.
- Business values created during a transformation that must be met for the business to survive. These values include retaining customers, producing relevant products or services, and creating revenue. They are described in ▶ Chap. 4.
- Environmental values are what are needed for people and businesses to operate with minimal disruption. These include the changing climate environment, which is described in ▶ Chap. 10, and liveability of cities, which is described in ▶ Chap. 11.

This chapter focuses on values held by people and the importance of designing transformations that add to these values.

### 3.1.1 Why Are Personal Values Important in Transformation?

Personal values are important for two reasons.

- Any transformation must result in improvement on the way services are delivered to people, and
- Activities in the organization to deliver services that meet people's values.

## 3.2 What Are Personal Values?

How do we define value—One way is what a person needs to be happy (Connors, 2017). Some examples include the following:

- Playing sport and being fit—so they value fitness. Which most people if not all people value.
- Having free time at home after a difficult workday is another value; maybe having more free time is the value provided by the meal delivery industry.

The common property of many of these is that their benefits cannot be measured financially. How do you put a number about people's attitude to fitness, or having free time at home? One way is to ask the following:

- What do you find important?
- What are you convinced that is true?
- What are the principles that you follow?
- What do you think of the way the business is currently working?
- How can things be done better?

### 3.2.1 What Is Important to You as a Person

Personal values are what a person values independent of the organization that they work in. They should be something you live and stand for. They should define you and what you stand for. Parents also try to instil these types of positive core values in children to give them guiding principles for living a good life. There are also values like not having to queue up for services or going through a sequence of checks in travelling.

Values in many cases are not easy to change—maintaining the status quo may lead to resistance to change, whereas thriftiness may result in change to reduce costs of a person's activities. Typical values include the following:

- Reducing waste and addressing climate change.
- Being a good steward of resources and in exercising frugality.
- Having good family relationships is of fundamental importance.
- Honesty is always the best policy and that trust must be earned.
- Maintaining a healthy work/life balance.
- Having pets and looking after them.
- Satisfying society goals.
- Completing projects on time and to budget.

Then of course there are values that people might have about a business, community, or network. For example:

- A business should maximize its contribution to society and not its own values.
- A business should maximize its reserve.
- Employees should be free to choose how they accomplish their tasks.
- There must be strict adherence to rules.

### 3.2.2 What Principles Do You Follow?

There are also your own personal feelings. Examples are shown in ◘ Table 3.1.

### 3.2.3 What Are You Convinced About?

People's strong convictions also play a role in finding values and finding ways to satisfy them. For example, vegetarians may be convinced that not eating meat is good for both health reasons and animal welfare. In that case, a restaurant owner might want to include vegetarian dishes in the restaurant menu to attract vegetarians.

A more relevant case is living in a city that is free from disease. In that people will agree that locking down reduces transmission of disease. But then it conflicts with another value, such as visiting family or going shopping. Moving around is necessary. Good design must achieve a balance of such values, such as restricting number of people in a venue.

3

Table 3.1 Personal principles

| Personal values | What to do to create value |
|---|---|
| I can inspire people | Any idea that inspires others |
| I like to contribute to communities and provide service to others | Make positive contribution to your community |
| I want to be healthy and fit | Develop activities to help improve health. |
| I see sport as important | Develop sporting venues and supporting sports clubs |
| I have good family relationships | Maintain good family relationships |
| I love nature and have pets | Go hiking, study nature, and take my dog for a walk |
| I am creative | Always doing something new and exciting that creates value for others |
| I can work in complex environments | Be able to explain complex systems in simple language |
| I am successful at what I do | Achieving your goals |
| I can build productive relationships | Produce some object with others. |
| I use my time productively | Always get value for your time. |
| I want to be mobile | Travel to other places. |
| I am thrifty | Choose minimal cost products |
| I like to share knowledge with others | Have frequent meetings |
| I accept change | Accept change if values are improved |

Negative attitudes and core values can also develop when people live in fear or insecurity and are forced to focus on survival in difficult circumstances Some examples of negative values include a feeling that:
- The world is a fundamentally brutal place and that only the strong survive.
- People are powerless to change their fates or personal situations.
- You do not deserve good things or relationships in life.
- Other people are fundamentally untrustworthy and unloving.
- Life is meaningless.
- Climate change will lead to continuing natural disasters.
- Work will be difficult to find.

## 3.3 How to Find Personal Values?

It is important to find the values of people, who will be affected by a transformation, have before starting a transformation. Interviews are a common practice here.

### 3.3.1 What to Find Out by Interviews?

While collecting information, designers usually record what they collect in documents that can be used later to make informed decisions. In addition to recording values, you should also find out what they do, how they work, and so on. Also record any suggestions they have or uses they see.

### 3.3.2 Using Ideas from Design Thinking

Design thinking focuses on finding out in depth issues through empathy and questioning of stakeholders. Initially, it is sometimes useful to use questions that focus on what people do and think. These develop empathy (Ross, 2019) with stakeholders, followed by collaborative brainstorming to identify their needs and concerns, and suggestions of ways to address any issues and problems.

## 3.4 Where Are Personal Values Important?

Values are often seen as a driving factor to motivate people to contribute more to the transformation whether it is in business or a community. It applies both in private and public life. Most people would argue that employees perform better if they see that the transformation achieves social goals and leads to business success.

The previous chapter outlined that collaboration is needed to transform in complex environments. Such collaboration is needed to motivate people to address core values, and to address conflicts between stakeholders in any transformation. A close match of personal and organizational values can therefore improve motivation and people's contribution to their organization.

Values are important in the following:

- Fitting employees into organizations.
- Developing communities.
- Delivering value to customers.
- Making organizational changes.
- Fostering collaboration and innovation.

### 3.4.1 Fitting Employees Into Organizations

Businesses and organizations need people not only with a good skill set but also those whose values align to those of their organization. For example, hiring a person who does not believe in following well-defined processes may not be the best for the construction industry.

### 3.4.2 Developing Communities

People with similar value often join to form communities such as local clubs, youth groups, women's groups, sports clubs, volunteer groups, and so on (Matinheikki, 2017). An annual conference on World Values Day challenges us to think about our own values and the values held by the groups we belong to and encourages us to act on those values. It stresses that using values in our activities ultimately can change our communities and the world we live in for the better (see reference for World Values Day in Further Reading).



The consensus is that more effective communities have shared values. Shared values are needed to work towards a common goal. Businesses foster shared values through staff motivation and reward systems. Responses to disruption need similar motivation to develop shared values. Cities need to develop values where health and safety is paramount.

### 3.4.3 Delivering Value to Customers

McDonalds is an example of a business that started small and became a global enterprise by delivering value initially to travellers but growing to serve people who need meals delivered quickly. It initially attributed its success preparing meals for travelling truck drivers through California who required minimal wait times. Reducing time waiting for meals was identified as an important value to drivers. Who were often annoyed when orders took time to prepare? The MacDonald brothers reduced waiting time by providing a quick but tasty combination that satisfied truck drivers. Then it turned out that it also met values of other buyers who are in a hurry and want a quick meal.

Over time, businesses adapt to changing values to grow. If you look at MacDonald's today, it is totally different than what was a hamburger outlet at the start. The value created was to reduce the travel time for long-distance drivers. Now McDonald's services include vegetable servings, coffee, and so on. But the basic value creation is still the same—quick service.

### 3.4.4 Making an Organizational Transformation

Changes in the way organizations work in most cases impact personal values. Most changes impact people values, and often lead to conflicts if some people see impact on their values as negative, while there is positive impact on others. In any transformation, it is important to ensure that value changes are acceptable to all. ◘ Figure 3.2 illustrates the impact of transformation on the values of different stakeholders. Some change adding to values of some stakeholders and not others, while other changes may have the opposite effect. Conflicts arise where changes are seen to add value to one and not another, sometimes resulting in conflicts on what change to accept. ► Chapter 7 describes the way conflicts can be resolved.

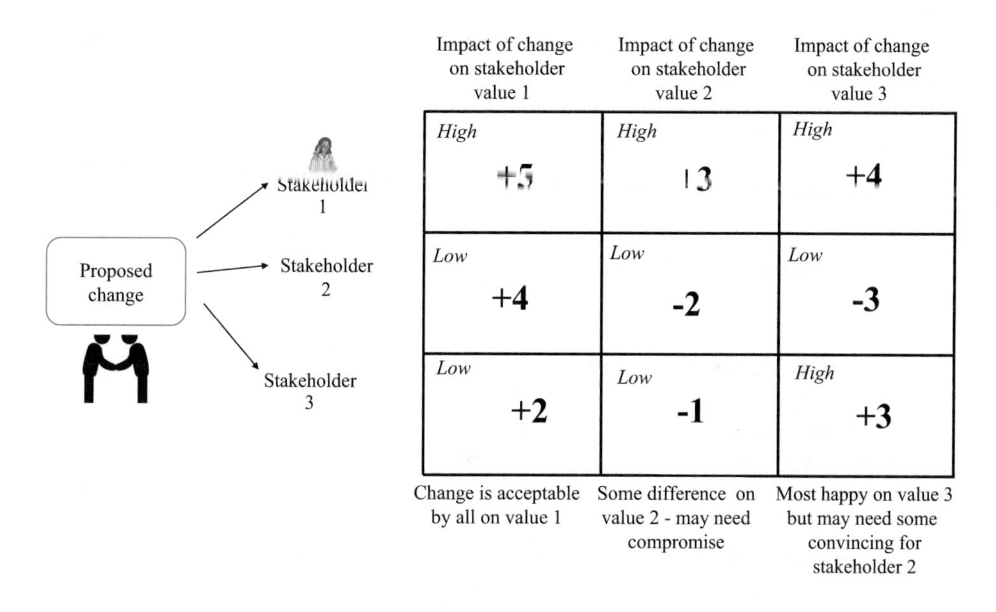


**Fig. 3.2** The impact of transformation on stakeholder values

## 3.4.5 Fostering Collaboration and Innovation

Continual change calls for any organization to continually innovate and remain relevant by developing products and services that deliver value to its customers. Hence organizations need people who value new ways of doing things. Should new ideas emerge informally or through new organizational structures and collaborative ways of working, as described earlier in ▶ Chap. 2. Perhaps to answer how to encourage innovation we need to define innovation itself. To some people, innovation is some idea that results in financial benefits. To others, it is a continuous process of product and process improvement.

We are often amazed at innovations such as Apple. Initially the intention was to deliver music, but where is it now? These are often the result of many years' work, including convincing the music industry to sell individual songs electronically Then, once electronically sharing became accepted, new features were added—camera to mobile phone is one example. A considerable number of such features often result from customer suggestions.

A series of articles from the MIT Sloan Management Review recently addressed the importance of innovation within organizations in an environment of continual change. Typical suggestions are to work smarter not harder or improve the rhythm of collaboration.

The consensus is one where multidisciplinary teams brainstorm ideas and ways they would add to customer value. Existing literature identifies knowledge sharing and developing trust significantly influence knowledge-sharing behaviour in addition to personal factors, such as knowledge self-efficacy and enjoyment in helping others.

Multidisciplinary teams, described earlier in ▶ Chap. 2, should include potential key customers. One question here is what the best way for such teams is to operate.

Should they be brainstorming continually, be done in open meetings, or through collaborative technologies. The answer probably is a combination, but in a systematic manner.

What is a good balance of activities in innovation? Maybe we need the equivalent of Fitbit to measure how much time people spend on different kinds of collaboration in an innovative environment.



Later, ▶ Chap. 5 describes different processes—the Stanford approach or the Double Diamond method as ways to organize collaborative work.

## 3.5 Recording What You Found

After interviews and discussions with people in the organization, it is necessary to record the outcomes. In design thinking, values are often shown on persona or empathy maps. One important criterion on persona maps is what are the major pains and gains.

The term empathy map is often used to characterize an individual. A persona map shows the characteristics of a similar group of people. For example, there can be a persona map for customers, tourists, or managers.

◘ Figure 3.3 shows a persona map that records information about customers—a typical customer. Do not develop a persona map for each person. A persona map is

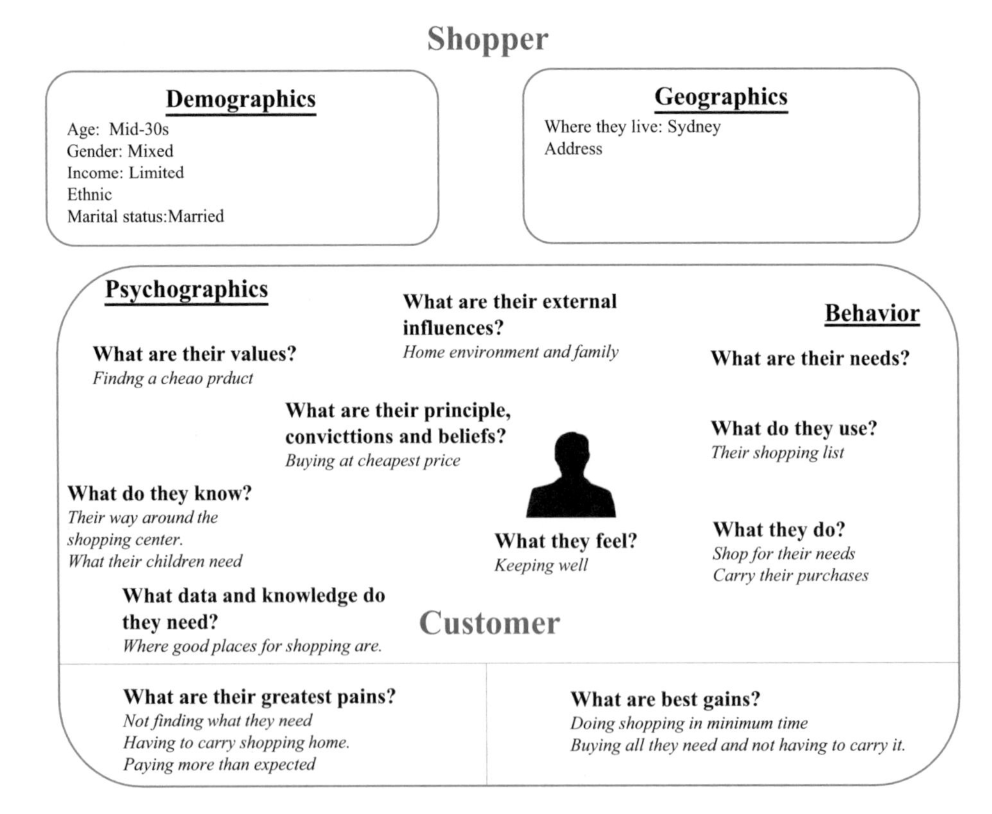


◘ **Fig. 3.3** Persona map

for people who share common characteristics. Thus, there will be a persona map for elderly customers, another one for younger, and so on. ▣ Figure 3.3 is one example of a persona map; there are many others that you can find by using Google. ▣ Figure 3.3 is an example of a persona map that shows characteristics of shoppers in a shopping mall.

Persona maps, or what are sometimes called empathy maps, are now commonly used in industry. They bring together information you have gathered about people and are used later in designing transformations that satisfy people's values and help then to carry out activities in better ways. There are a large variety of ways to diagrammatically show persona maps. ▣ Table 3.2, for example, shows a tabular presentation. You might notice two sections in ▣ Fig. 3.3; what are their greatest pains, and what are their greatest gains. These two questions appear in virtually any persona map. These play almost a central role in making design choices.

▣ **Table 3.2** A tabular persona map

| **PERSONA MAP**<br>**PERSONA NAME: ___ CUSTOMERS: ________________**<br>**DEMOGRAPHICS: Age: Mid-30s___ Income: Limited ____ Other: _________**<br>**GEOGRAPHICS.**<br>**Location, Country: Sydney, Australia Organization:** | |
|---|---|
| **Themes** | **Description** |
| Convictions, principles, and beliefs | I like to pay the minimum price<br>My time is valuable<br>Climate change is happening |
| What are their needs? | Ways to find the cheapest price for a product<br>A shopping list |
| Knowledge | |
| What they know? | Their way around the shopping centre<br>What their families need |
| What do they need to know? | Their way around the shopping centre |
| What would they like to know? | The best shopping places |
| What persona they frequently communicate with? | Shop keepers |
| Ability | |
| What do they do? | Negotiate price with the seller<br>Buying at the cheapest price |
| What would they like to do? | Shop for their needs<br>Carry their purchases to the car |
| What are their greatest pains? | Not finding what they need<br>Paying more than expected |
| What are their greatest gains? | Doing shopping in minimum time<br>Buying what they need and not having to carry it<br>Paying the lowest price |

In this book are sections on values, needs, and knowledge. All of these will have an impact on shopper behaviour and what they buy. For example, addressing climate change can influence what people buy, as their decisions may be influenced by the emissions created when manufacturing a product.



### 3.5.1 Identifying System Components from Interviews and Reports

When interviewing stakeholders, it is useful to identify components that stakeholders describe. So, from interviews you can find out where people work, what they do, what they think about the various parts of the organization, and so on.

Thus, as stories emerge, it is sometimes worthwhile to develop a system model. ◘ Figure 3.4 is a simple example of an emerging model that initially appear as sketches. Later, in ▶ Chap. 6, these will be put together into a model of the system.

For example:

- We start with putting stakeholders in the middle.
- Then we add where they work.
- Then what do they do, and so on.

While documenting responses, you might make sketches such as those shown in ◘ Fig. 3.4. In ▶ Chap. 6, these become part of an organizational model.

It is important to continually ask questions about what is going on in the system and add to the model. For example, what are the data needs, what processing is needed, and so on.

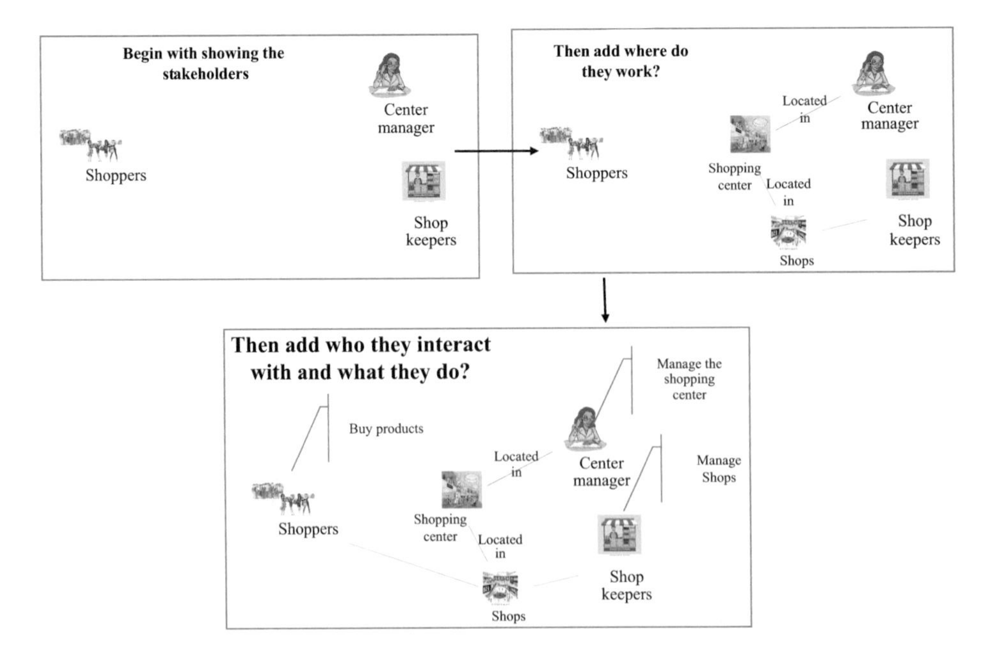


◘ **Fig. 3.4** Sketching infrastructure

## 3.6 Customer Segmentation

It is generally agreed that one important persona to any business is the customer (Landoguez, 2013). One of the most important things in analysis is to identify your customers and their values. ◻ Figure 3.3 is a persona map that described characteristics of a typical shopper. In most businesses and cities, there are many different customers. These different kinds of customers are called customer segments. Customer segmentation is commonly used in design to classify business customers. Any business now knows that different customers have different values and hence may require products and services that satisfy these values. Different ways are needed to:

- Organize the support of key customers.
- Gain alignment and address conflicts as early as possible.

Often, in analysis, we identify similarities and differences between customer segments. There is a different persona map for each segment. Differences in customer segments can also be recorded in tables such as ◻ Table 3.3, which shows some examples of different customer segments.

## 3.7 Who Are the Stakeholders?

The word "stakeholder" is now playing a dominant role in businesses and society. In a business, it may be an investor who set up a company, a shareholder, business executives, and so on. It is also now becoming increasingly common to talk about stakeholder engagement. Engaged stakeholders are highly likely to contribute to transformations that meet their values. Thus, citizens should be engaged in developing city services, that add value to the citizens as well as making the city more attractive. Customers of a business should be engaged to contribute ideas to improve products and services of a business.

One of the first steps in any transformation is to identify the stakeholders who will be impacted by any new development. Any such new developments create value

◻ **Table 3.3** Showing customer segments

| Customer segment | Family group | Young person | Elderly | Professional office worker |
|---|---|---|---|---|
| What they buy | Highly likely to buy toys, shop for food, especially sweets and drinks | Clothing and food | Food, medicine | Stationary, office clothing |
| Greatest pain | Unhappy children | Finding everything very expensive | Not finding places to sit down | Not being able to find things quickly |
| Greatest gain | Finding activities for children | Getting cheap fashionable clothing | Easy ways to get between shops | Developing new contacts |

for these stakeholders. Value creation is more likely to happen if they are engaged in the transformation. What do we need to do to engage stakeholders? The first steps are to:
- identify who the stakeholders are.
- segment them.
- rank in importance.



For example, someone may be an employee, but also in terms of their responsibility—their role in the organization, or their responsibility and what they do. It is also possible for a stakeholder to have more than one role.

There is a variety of stakeholders in each network. Some are industry specific, others are general. General stakeholders are those that are found in most design environments. They may not be involved directly in the project.

### 3.7.1 Stakeholders Found in Most Organizations

Table 3.4 shows the kind of stakeholders found in most organizations.

Other parameters often used in stakeholder analysis include the following:
- The impact of change of a system on a stakeholder.
- How much influence they have on decisions.
- What is important to them?
- Their contribution to the new system.
- Can they block the design or parts of the design?

### 3.7.2 Industry-Specific Stakeholders

These are stakeholders that are specific to an industry, for example, Table 3.5 shows stakeholders typically found in education.

**Table 3.4** Stakeholders found in most organizations

| Stakeholder | Values |
|---|---|
| Employees | Will the change improve the organization and make my position more secure? |
| Management | What will be the effect on my investment or section? |
| Suppliers | Will deliveries be constant and growing? |
| Customers | What are the customer needs? |
| Investors | Will the change add to business value? |
| Government | Setting industry regulation. |
| Local communities | What will be the impact of any change on the community? |
| Owners | What will be the impact on assets? |

■ **Table 3.5** Industry(education)-focused stakeholders

| Stake-holder | What they do that has impact | What they value | Ways to satisfy value |
|---|---|---|---|
| Teachers | Create learning materials. Teach classes | Good teaching materials. Good teaching outcomes. | There is a large difference in students in class |
| Policy-makers | Allocate resources. Change what teachers do. | Better learning outcomes. Reduced cost. | Define projects to improve learning. |
| Students | Attend school. Change schools. | Good learning materials and teaching methods. | |

■ **Table 3.6** Stakeholder importance values

| Stake-holder | Impact of change on stakeholder | Influence on project | What they contribute | Do they have veto power? |
|---|---|---|---|---|
| Investor | Must be informed of any change | Provides funding to project | Funding of project | Most probably |
| Project manager | Change of project schedules | Can provide leadership to finish faster | Leadership | Only of the way tasks are carried out |

### 3.7.3 Who Are Important Stakeholders?

There are also stakeholders that have more influence in the organization such as those in ■ Table 3.6. Often, values of such stakeholders also have higher importance.

## 3.8 Value and Transformation

A good transformation creates additional value for stakeholders. Each business often has their set of values—they often develop values to match those of their existing or potential customers. Customer values in what the organization sees as important. The success of their products or services will be greater if they add to customer value.

One essential aspect of good design is to create value for people in any system you are designing. Normally, the term value was used in a financial context, but can refer to many other things. Values now are often more than financial, they can, for example, be, love to travel, access to knowledge, or a clean environment. We do not often buy or eat the cheapest food; most people prefer to spend a bit more money to eat what they like or what is healthy—they sacrifice some financial value to add to health value.

### 3.8.1 Designing to Create Stakeholder Engagement

The development of knowledge proceeds in several steps.

- Step 1—Identify stakeholder roles and their responsibilities. Finding knowledge about stakeholders can be through quantitative surveys or through qualitative sessions with key people. The knowledge found is recorded on persona maps.
- Step 2—How many of each role. Analysis and quantification of values.
- Step 3—Identify what stakeholders do and what they need to carry out their activities.
- Step 4—Find out what knowledge they have and need.



For whom are you designing the system? Focusing on values in design presents several options.

- Focus on a specific stakeholder, often a customer
- Specific business or city value
- A combination of both.

**Summary**
This chapter described the wide range of values. It made a distinction between personal and stakeholder's values. Personal values are those that people have, irrespective of where they work or what organization they support. Any design should take such values into consideration. Designs that meet personal values result in greater commitment of people and consequently in better outcomes and organizational performance. Organizational values will be described in the next chapter.

**Exercises**

Now we go into more depth in the case studies introduced in ▶ Chap. 1:

- Identify their stakeholders and their values. Use a table as shown in ◘ Table 3.4 to provide a stakeholder list.
- Develop a persona map for select stakeholders.
- Draw some simple sketches as those in ◘ Fig. 3.4.

**Shopping Centre**

- Who are the stakeholders in a shopping centre?

**Restaurants**

- Who are the stakeholders in a typical restaurant?
- Develop a persona map for some selected stakeholders.
- Can you segment some of the persona?

**Food Supply Chain**

Look at the food supply chain you developed in ▶ Chap. 1.

- What are the major stakeholders of each step of the supply chain?
- What are their values?

**Your Personal Values**

Just for interest, ask yourself questions such as what is important to you, what are your convictions, and what principles you adhere to?

**Match with Others to Choose a Group You Can Work With**

Do not just make something up—provide evidence. For example:

- What is the last decision you made and what values did it address?
- When did you feel you did your best?
- What are you proud of?

## Further Reading

Connors, C. D. (2017). *The value of you: The guide to living boldly and joyfully through the power of core values*. Skull and Paddle.

Ditman, H. (2008). *Consumer culture, identity and well-being*. Psychology Press.

Landroguez, S. M., Castro, C. B., & Cepeda-Carrion, G. A. (2013). Developing an integrated vision of customer value. *Journal of Service Marketing, 27*(3), 234–244.

Matinheikki, J., Rajala, R., & Peltokorpi, A. (2017). From the profit of one toward benefitting many – Crafting a vision of shared value creation. *Journal of Cleaner Production, 162*, 583–591.

Noon, M., Blyton, P., & Morrell, K. (2013). *The realities of work*. Palgrave Macmillan.

Porter, M. E., Hills, G., Pfitzer, M., Parscheke, S., & Hawkins, E. (2013). *Measuring shared value: How to value by linking social and business results*. FSG.

Ross, M. (2019). *The empathy edge: Harnessing the value of compassion as an engine for success*. Two Page Books.

Spieth, P., Schneider, S., Clauss, T., & Eichenberg, D. (2020). Value drivers of social businesses: A business model perspective. *Long Range Planning, 52*(3), 427–444.

World Values Day. https://www.worldvaluesday.com/wp-content/uploads/2020/05/WVD-2020-Values-Guide-for-Community-Groups-final-3.pdf (Last accessed: March 5, 2021).

Xi, K., Wu, Y., Xiao, J., & Hu, Q. (2016). Value co-creation between firms and customers: The role of big data-based cooperative assets. *Information and Management, 53*, 1034–1048.

# Organizational and Business Values

Contents




I. T. Hawryszkiewycz, *Transforming Organizations in Disruptive Environments*,
https://doi.org/10.1007/978-981-16-1453-8_4

This chapter extends discussion on values to organizational and business values. Values are important because organizations and businesses, whose values are aligned to customer values, tend to be more successful.

This chapter describes the importance of alignment and how good alignment creates value. It then defines how value alignment is achieved by services that deliver what the organization and its businesses and stakeholders want. This chapter then describes how businesses operating in cities or other communities must work together to create an environment that creates value for both.



**Learning Objectives**

- Challenges faced by organizations
- The triple bottom line
- Core business values and missions
- Developing capabilities that realize values
- The organization as a system.

## 4.1 Introduction

This chapter extends discussion from personal to businesses values and the importance of aligning business values to personal values. People often join businesses because they have similar values. Organizational and business values define what organizations do to survive in today's complex environment, sometimes called their mission or vision. The business environment now is composed of businesses that range from large global conglomerates to small shops; there are also start-ups as well as community organizations. Whatever their size or vision, they all work in complex environments and must deliver value to their clients and to other businesses. At the same time, they must address the challenges in the complex environment and be good global citizens (Coletta, 2018; Dorst, 2016). Any transformation must find ways to align stakeholder values with organizational values, while satisfying environmental values.

For many years, businesses followed the traditional production chain, often called the Porter's value chain (Porter, 1980), where a business buys raw materials, adds value to them using some well-defined method, and then sells and realizes a profit. Traditional supply chains are now less common. Customer values are continually changing, their needs are changing as are ways in which services are delivered. A traditional value chain can no longer ëasily satisfy newly emerging challenges.

Within the context of globalization, it is the supply chain that became dominant. Here value is added as one proceeds through the supply chain. Getting products and services quickly to the point where they are needed became dominant—the just-in-time syndrome. New values now include resilience to disruption in the face of the global COVID-19 pandemic. Resilience is becoming increasingly dominant, especially in supply chains that must deliver goods and services during disruption.

### 4.1.1 Triple Bottom Line

The triple bottom line was introduced in ▶ Fig. 3.1, which is increasingly used as a broad guideline for organizational transformation. It now shifts the focus of design from purely commercial issues to include business contribution to addressing environmental and global issues such as climate change. Aligning business and environmental is now seen as a corporate responsibility to reduce the impact of climate change—something especially relevant to younger customer segments. ◘ Figure 4.1 extends ▶ Fig. 3.1 showing alignment of the three sets of values.

Emphasis on environmental values shows greater awareness of people to the possible natural disruptions resulting from climate change, and values of people in preserving the natural environment. Consequently, there is more emphasis on business to minimize their impact on the environment. How businesses can minimize their impact on the environment and make it part of business values will be described in more detail in ▶ Chap. 10. The goals here are as follows:

- Business activities do not contribute further to climate change and lead to further disruptions from the planet.
- Businesses operating in ways that can mitigate such disruption.

## 4.2 What Are the Emerging Challenges?

There is no theory that defines the challenges faced by organizations arising from the dynamic nature of the global environment. Many of these come from observations of practitioners, various industry reports, as well as academic surveys. Three examples are shown in ◘ Fig. 4.2. These are examples in pre-COVID days. One of the major challenges is uncertainties in the environment and how to deal with them (Teece, 2016).

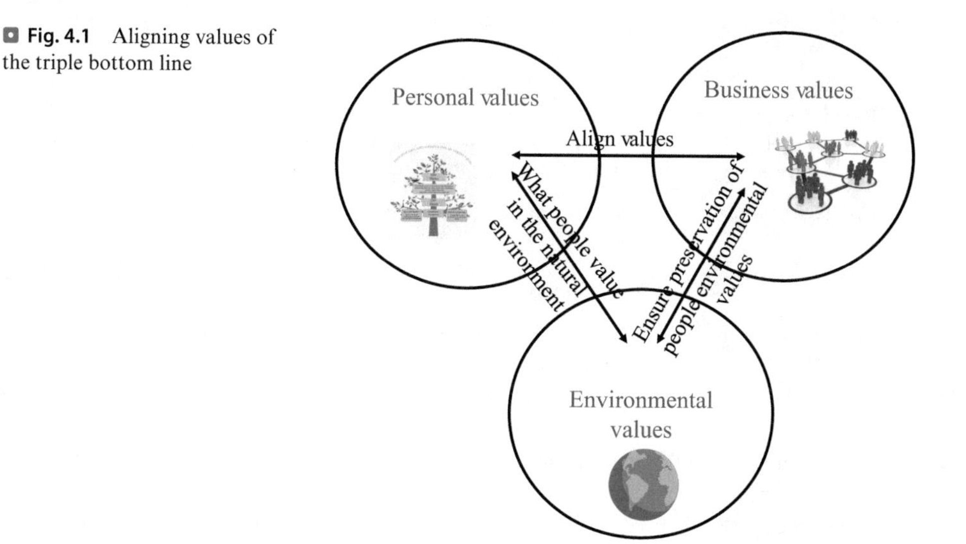


◘ Fig. 4.1 Aligning values of the triple bottom line

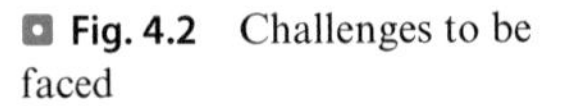

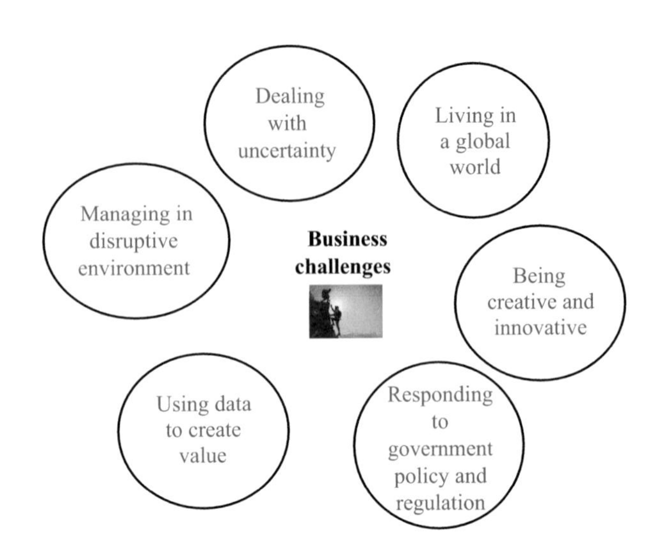


Fig. 4.2 Challenges to be faced



Table 4.1 Business challenges

| Business challenge themes | Theme descriptions |
|---|---|
| Dealing with uncertainty | Planning for longer term focus, but acting in short term. Long-term plans may include better predictive and innovation capabilities. Agility is a requirement here. |
| Living in a global world | Businesses must think globally and have products and services that can quickly enter any country.<br>Being part of global supply chains. |
| Being creative and innovative | To be competitive in a global business environment, businesses need to continually innovate and develop new products and services.<br>Ability to manage different viewpoints and resolve conflicts. |
| Responding to changes in government policy and regulation | Especially in industries such as energy, banking. Taxes are also important here, as is maintaining health of your employees. Interpretation of relevance to the business. |
| Using data to create value | Maintaining a flexible technical architecture and data analytics support. Making sense of data environment of information overload.<br>Dealing with an increased amount of information and its relevance to the business. |
| Managing in disruptive environment | Coordination between different parts of the supply chain. Dealing with disruptions. Ensuring healthy environment. |

The emerging challenges shown in Fig. 4.2 and Table 4.1 can be derived by analysing of industry reports such as:

- Lean Methods Group, a global firm specializing in business problems has Top Ten Problems Faced by Business (see references) that include uncertainty, globalization, innovation, and government regulation.

- Hiscox—see the ten biggest challenges businesses face today (and need consultants for) and uncertainty about the future.
- Forbes—The "8 Great" Challenges Every Business Faces (And How To Master Them All) that also include uncertainty and regulation.

▪ Table 4.1 is an overview of challenges found in these and other reports. The most common one is the uncertainty in the environment and how to deal with it. Businesses are now required to be more agile and resilient. Government regulation as well as technology also are common.

Changes of government regulation or emerging technologies can be seen as potential disruptors. How then do you design a business to address these uncertainties? Response to major disruption calls for major changes in business and society as evidenced in the COVID-19 pandemic. These are discussed later in ▶ Chap. 9 and often result in a major realignment of values caused by organizational or environmental change.

Most organizations now go further with their mission statements and now include environmental issues as part of their values, as demonstrated by the triple bottom line.

## 4.3 Organization Values

Most organizations now see themselves as being socially responsible while being commercially successful. At the same time, guided by the triple bottom line, a third dimension is emerging, resilience and sustainability. Resilience will be described in more detail in ▶ Chap. 9.

### 4.3.1 Core Social Values

Organizations often have what are called core values. These are what are "good" in the community. There are no standard core values, but their focus is on being good citizens. Some values are illustrated in ▪ Fig. 4.3 to give a general overview of the kind of core values that are found in many organizations. They define the themes (shown as circles) to be addressed in any transformation.

In summary, core values for an organization can include the themes shown in ▪ Table 4.2.

Two common characteristics of corporate visions are that they look to the future and address future needs of people. For example, *Google's* corporate *mission* is "to organize the world's information and make it universally accessible and useful". The *mission* of Massachusetts Institute of Technology (MIT) is to advance knowledge and educate students in science, technology, and other areas of scholarship that will best serve the nation and the world in the twenty-first century. Other examples include organizations such as the following:

1. Apple Computer is perhaps best known for having a commitment to innovation as a core value. Such commitment to innovation is embodied by their "Think Different" motto.

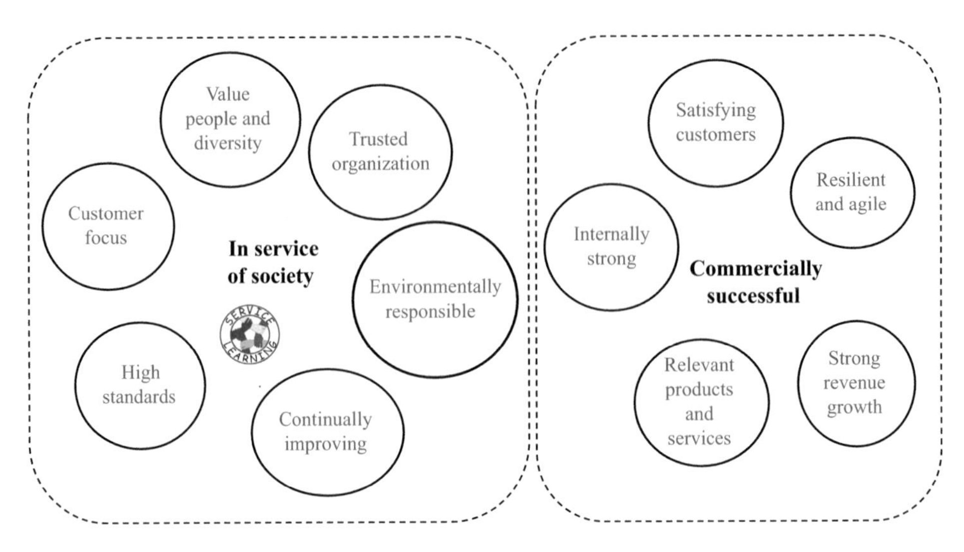


Fig. 4.3 Organizational values and Themes

Table 4.2 Social themes

| Social themes | Description |
|---|---|
| Customer focus | Being agile to address customers' changing values as, for example, a trend to healthy food<br>Building strong communities<br>Being creative and introducing new experiences |
| High standards | Striving for leadership by leading-edge products and services |
| Continually improving | Being agile to address customers' changing values as, for example, a trend to healthy food<br>Being innovative to respond to new challenges |
| Being a trusted organization | Being dependable on meeting their customer needs<br>Helpful in emergencies and disasters |
| Environmentally responsible | A commitment to sustainability and to acting in an environmentally friendly way |
| Value people diversity | Showing concern for people and compassionate.<br>A commitment for helping people less fortunate |

2. Royal Dutch Shell Oil company donates millions of dollars to the University of Texas to improve student education and to match employee charitable donations.
3. Patagonia, an outdoor clothing manufacturer, whose core values include using business to protect nature, cause no unnecessary harm, while making the best products (see references).
4. Ben & Jerry's, while making fantastic ice cream as a goal, have environmental sustainability as a core value, their vision is to make the world a better place and create sustainable financial growth.

It is also important to recognize that people and business values can change over time. People may develop new preferences; organizations change services and products; and they may also enter new businesses. Emphasis on minimizing impact on the environment is one example here.

Most organizations now publish their values, often as a mission statement. Apart from having values that are generally seen as a contribution to society in general, most organizations have a specialized mission that states how they add to society value.

### 4.3.2 Commercial Values

Commercial values are those that make a business commercially viable. The focus on commercial values is to grow customers. The four major values to build a customer base are as follows:

- Maintaining customer satisfaction
- Developing relevant products and services that customers need
- Maintaining a strong internal environment to produce products in the best possible way or deliver services and
- Building a revenue base through a growing list of customers.

### 4.3.3 Organizational Missions

Business is now becoming more specialized. They must transform to an internal structure that can deal with emerging challenges and increasingly changing customer needs. We use the term "challenges" rather than "problems", as often the problems themselves need to be defined. Businesses must develop the capabilities to address new challenges. The needed capabilities are roughly described in ▫ Fig. 4.4. It includes the internal environment such as human resources or plant and machinery. But increasingly it is necessary to consider the external environment in design.

## 4.4 Developing Capabilities to Realize Values

The goal of any transformation is to create organizations that realize the values. They must ensure that organizations have the capabilities to do so. ▫ Figure 4.4 illustrates some capabilities.

One important goal of design is aligning business and employee values. If business and client values are not aligned, then it is more likely that you will find it hard to keep your clients. If values are not aligned, then a business may be producing items and services that have little value to your clients. Similarly, your business employee values should be aligned to your business values as they will feel more motivated in their work.

Many of the challenges in ▫ Table 4.1 cannot be influenced by a single business. Complexity or population growth are factors that an individual business cannot

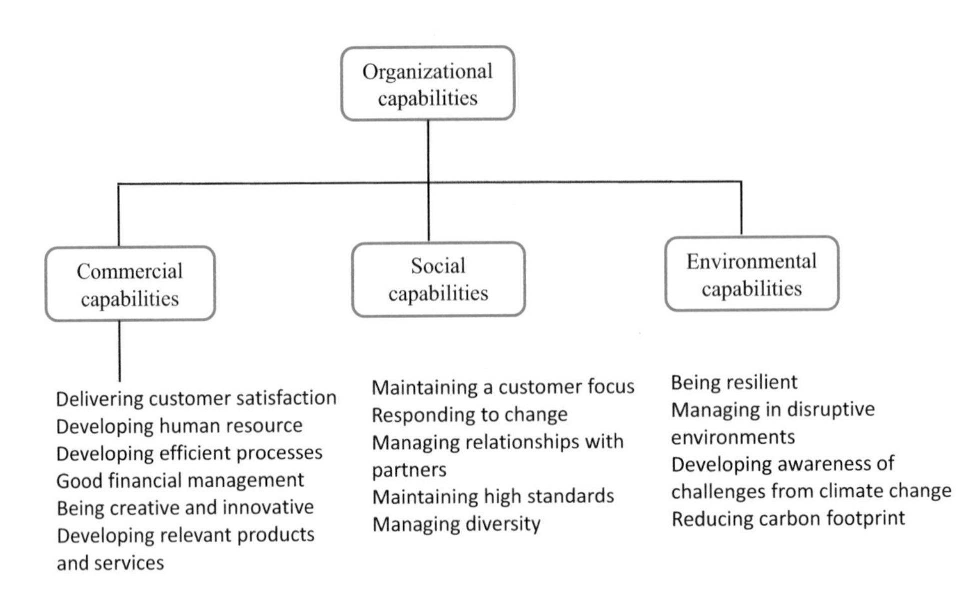


**Fig. 4.4** Organizational capabilities

influence. Another is government regulation. Uncertainty, complexity, and information overload are inherent in the environment and must be accepted as normal in design. Some challenges are those which you can hope to influence in your design.

### 4.4.1 Commercial Capabilities

These are capabilities that a business can and needs to develop to be competitive. They are summarized in Table 4.3 and are drawn from empirical evidence such as that in Fig. 4.3. The summary identifies four capabilities and some examples of what can be done to develop them.

Developing necessary capabilities are the design challenges, something that designers can influence and change. They should become part of the values held by employees.

### 4.4.2 Capabilities for Dealing with Complexity and Disruption

So far, we have identified what may be called traditional business values. In the past, a design of business focused on business criteria primarily on ultimate financial success. The focus was usually on customers' segments and matching them to products and services. Increasing complexity now places many new challenges. These are to be adaptive to new competitors, evolution of technology and human relationships. Businesses must develop capabilities to deal with such disruption. Two new capabilities here are resilience and agility. Some are shown in Table 4.4 and described in more detail in ▶ Chap. 9.

**Table 4.3** Commercial capabilities

| Business value themes | What capabilities are needed to address the theme |
|---|---|
| Maintaining and growing a customer base through improved customer satisfaction | Developing customer loyalty program<br>Maintaining customer channels<br>Developing ways to grow customers<br>Protecting customer safety and health<br>Continually improving products and services |
| Building a revenue base | Identifying opportunities where we can expand our offerings |
| Being relevant by being part of the development mainstream | Maintaining brand and reputation<br>Being aware of changing customer needs<br>Ensuring products and services continually meet changing needs |
| Maintaining a strong internal environment | Hiring policy to develop well-functioning and highly performing teams<br>Developing a safe environment<br>Supporting creativity and innovation |

**Table 4.4** Environmental value themes

| Environmental value themes | What capabilities are needed to address the theme |
|---|---|
| Business resilience | Forecasting to identify potential threats and opportunities<br>Anticipating and responding to change by developing new goals and directions |
| Agility | Being able to change business processes quickly. Being able to reassign people to new business activities quickly and reassign people to the new activities |
| Waste management | Making sure waste is disposed or recycled<br>Reducing emissions from waste disposal |
| Energy conservation | Minimizing energy use<br>Using renewable energy whenever possible |
| Emission standards | Minimizing greenhouse gas emissions |

### 4.4.3 Addressing Climate Change

Now we address the third part of the triple bottom line—dealing with environmental values. Environmental issues, especially climate change, are now beginning to play a more significant role in decision-making.

The response to climate change is increasingly complex and is discussed in detail in later chapters. The response centres both on how to reduce emission of greenhouse gases, ways to respond to natural disasters, and increasingly on mitigation to reduce the damage. These are discussed in detail in ▶ Chap. 11.

## 4.5 How Are Capabilities Developed?



Transformations must add to values and address challenges that businesses face. One way to transform businesses in a systematic manner is to define them as a set of parts Then define how each part will be changed and how they will work together.

### 4.5.1 What Is a Business System?

Most people see systems from the engineering perspective. A car, for example, is composed of technical parts, the engine, a cabin, and the transmission system from the engine to the wheels. All these parts can be designed independently but, when put together, they work together as a moving car. Similarly business systems are made of parts that must work together. In this case, however, it is people that make the parts work together.

▫ Figure 4.5 is the model of the hotel as a set of systems. Each rectangular part is a system and ultimately, we develop what is commonly called a system of systems.

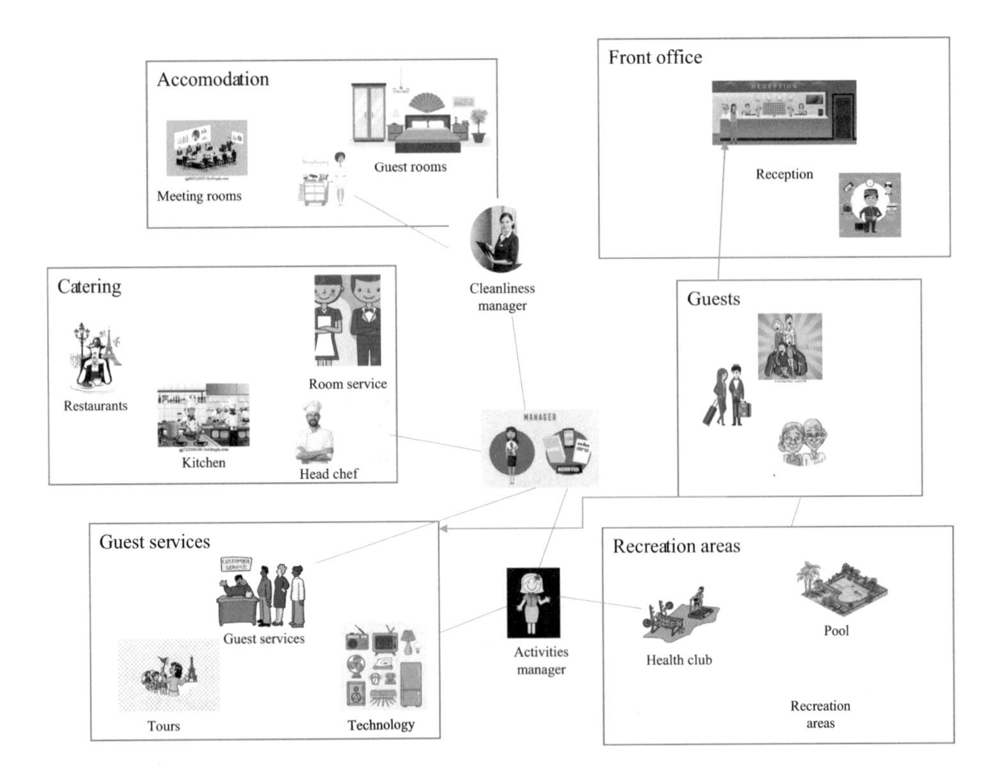


▫ **Fig. 4.5** The hotel as a system of systems

Figure 4.5 does not include all systems in a hotel, but shows typical systems such as accommodation, catering, recreation areas, the front office, and so on.

Now suppose we see the system from the perspective of delivering services to stakeholders. The term “system” now takes a new interdisciplinary meaning—with human and technical components, one is a technical part that behaves predictably to an input. The other, the human part, can behave unpredictably given an output from the technical part. In fact, most businesses can be described as a set of systems.

There are many other ways to show complex systems. These range from broad sketches to formal diagrams like that of Fig. 4.5 that show all connections. Rich pictures are a mix of these. In many cases, diagrams are there to show the design environment as a system. Ways of modelling systems are covered in more detail in ▶ Chap. 6.

### 4.5.2 Infrastructure to Support Services

Services delivery requires technology and physical structures. These are then used to provide services. Registering a new guest requires the front office a computer system to manage conferences, register guests, or arrange a tour. Others may be more complex such as arranging a conference. It also requires physical structures such as rooms and elevators to get people to rooms. All these physical components are called an infrastructure. Transformations often require a change in parts of the existing infrastructure or add new parts.

### 4.5.3 Developing a Platform to Provide Services

In Fig. 4.6, one separation would be by what guests value—some want clean individual rooms, others want to run conferences, still others may want local entertainment. We then need to develop platforms where hotel employees can use the infrastructure to deliver services. We ultimately create a system of systems that share a technology platform as shown in Fig. 4.6.

Figure 4.6 also shows the growing importance of technology as providing the devices to tie all the parts together and support the communication required between the services and their users.

Part of the infrastructure is a technology platform where people from different systems can see what everyone else is doing and contribute whenever possible to the global performance of the whole system.

*A guiding principle*—divides into services and infrastructure. Look at complexity as defining services to satisfy stakeholder needs. Then as evaluating how infrastructure can support the services.

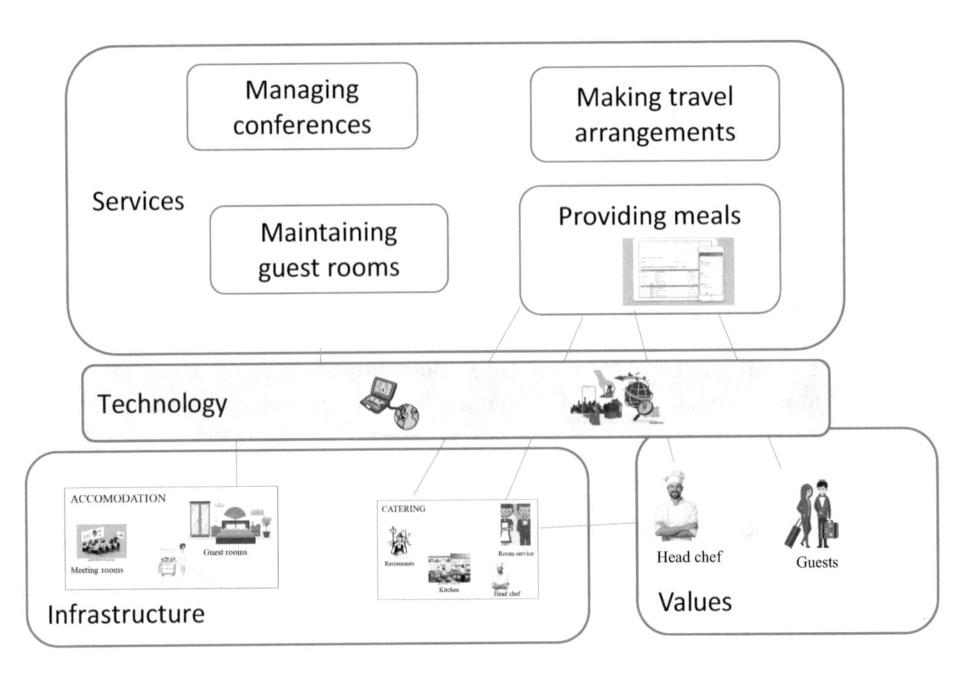


**Fig. 4.6** Infrastructure for services

Running systems independently often creates idle resources. To minimize resources being idle, all systems should be accessible across a platform that resources to be shared among systems.

Coordination between businesses is becoming important today with increasing emphasis on outsourcing of tasks in a complex process. Many organizations now outsource individual work from a process to different contractors. These tasks must be coordinated to achieve a good outcome in the shortest possible time. Often such coordination takes place through e-mail rather than a platform resulting in non-productive use of time.

## 4.6 The Role of Cities

Cities are important for organizations and their businesses as most carry out their business activities in cities and communities. Cities provide the infrastructure in which businesses work. This infrastructure includes traffic management, safety, venues for entertainment, among others.

Questions that can be asked here are "How can cities provide the infrastructure that adds value to businesses?" and "How can business carry out their activities while at the same time adding value to cities?" This question is very current at the time of writing. Cities have been hard hit by COVID-19 due to lockouts that meant few people use business services, especially hospitality services. Cities are in the process of recovering. To do so, they must work together with city businesses to develop guidelines for businesses to operate while adhering to COVID-SAFE rules that are set by cities.

The design processes that lead to creation of city infrastructure are described in detail in ▶ Chap. 11, city values and the way they add to business values.

### 4.6.1 City Environment

Just like businesses, each city environment is different. Each city is at a different level of developments and needs to address different issues. Lee and others (2014), for example, describe differences between the strategies adopted in Seoul and San Francisco.

### 4.6.2 Challenges in City Design

Design of cities itself must address challenges. Some of these are illustrated in ◘ Fig. 4.7 and ◘ Table 4.5. They are described in more detail in ▶ Chap. 11. What is important in this chapter is that cities must provide the services needed by businesses. The term often used here is that cities must provide the infrastructure to support businesses.

### 4.6.3 City Values

Most businesses now serve citizens in smart cities. These put additional criteria on business process design. Increasingly, business is being created to provide services to city dwellers while observing city rules and regulations.

Similarly, city dwellers and administrators define city values through various consultative arrangements. Each city may have different values and be at a different stage of development. ◘ Table 4.6 describes some of the more important city values.

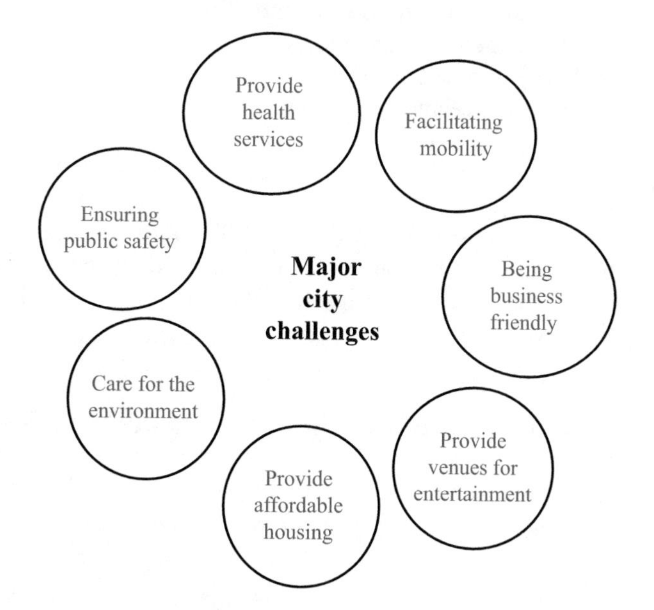


◘ **Fig. 4.7** City challenges

Table 4.5 City challenges

| City challenges | What are success measures? |
|---|---|
| Provide health services | Healthier citizens<br>Easy access to health services requiring less health services |
| Ensure public safety | Crime reduction<br>Quick response to disasters<br>Reduce spread of disease |
| Facilitate mobility | Improving traffic flow. Getting as quickly as possible from place to place |
| Be business friendly | New job being created<br>Provide facilities for business and social activities |
| Caring for the environment | Clean air, green areas<br>Reduced energy use |
| Provide affordable housing | Making housing affordable through city planning rules |
| Support education | Access to knowledge specially to upskill employees |
| Provide venues for entertainment | Available venues for sports, cinemas, theatres, and open areas |



One goal for identifying issues is to create better alignment of city infrastructure and services to community and business needs. This has been illustrated during the COVID-19 pandemic. Cities have been both highly impacted by the pandemic as they have been proving ways to spread the virus. In response, they have responded by emphasizing safety and health as critical and focusing on ways to prevent virus spread by lockdowns and social distancing, often creating value conflicts.

Just like Fig. 4.5, the city can also be a system of systems. For example,

- Transportation needed for mobility
- Smart energy, water, and waste
- Waste disposal for health
- Governance and education
- Safety management

Among others.

A city must develop the capabilities for such systems to deliver services to its inhabitants and businesses. Businesses then use city services. For example, the catering system in Fig. 4.5 would require waste disposal services provided by the city.

### 4.6.4 Data and Data Analytics to Make Cities Smart

Increasingly data analytics are important in creating smart cities. Examples include the following:

**Table 4.6** City values

| City value themes | What is needed to realize values |
|---|---|
| Walkability | Wide streets and open spaces<br>Preference to walking rather than cars<br>Places to rest or take breaks or meet people—coffee<br>Things to see while walking<br>Close to public transport |
| Safety | Low crime levels<br>Disaster preparedness<br>Low pollution levels<br>A safe city policy |
| Health | Services for citizen health<br>Good sanitation<br>Safe food<br>Access to clean water<br>Access to good food |
| Environment | Feeling part of nature<br>Shared outdoor spaces that promote physical activity<br>Increasing number of trees |
| Mobility | Easy ways to move from place to place<br>Convenient public transport services |
| Connectivity | How easy is it to connect with others? |
| Vibrant spaces | Community spaces<br>Encourage innovation |
| Sustainability | Sharing resources (public transportation)<br>Alternate energy sources increasing |

- Improving mobility through city transportation.
- Improving safety through identification of safety risks.
- Identifying needs of vulnerable citizens.
- Encouraging idea generation for community improvement.

One of the important questions here is whether any data collected about activities in a city are available to only one application or open to all.

**Summary**

This chapter described organizational values and challenges. It began with business values—values that businesses must meet to survive. It then introduced the triple bottom line. The triple bottom line provides additional values that businesses contribute to society in general. In our case, businesses usually operate in cities and hence any business change must contribute to city values. Increasingly, there is awareness of the effects of climate change and that business changes should reduce the dangers associated with climate change.

4

### Exercises

For each of the case studies introduced in ▶ Chap. 1, do the following.

Identify the systems that make up the organization and draw a system diagram like ◘ Fig. 4.6. For each system, identify the value it will create for the business. The values are defined in ▶ Sect. 4.3.2. Identify the services each system will provide to stakeholders.

In the restaurant, they can be kitchen, service counter, customer dining. In a shopping centre, they can be car park, shops, rest areas, food courts, and so on.

In the food supply chain, they can be farms, factories, supermarkets, restaurants. Each of these systems has its own values but also must satisfy the values of other systems that have their values. Describe the kinds of businesses that are part of the supply chain. Identify potential disruptions to the supply chain. Distinguish between disruptions to the business, and transfers of goods between them.

Use the themes in ◘ Table 4.1 to identify the challenges faced in each case. Then describe the business capabilities to address these challenges.

## References

Coletta, C., Evans, L., Heaphy, L., & Kitchin, R. (2018). *Creating smart cities*. Routledge.

Dorst, K., Kaldor, L., Klippan, L., & Watson, R. (2016). *Designing for the common good*. BIS Publishers.

Lee, J. H., Hancock, M. G., & Hu, M.-C. (2014). Towards an effective framework for building smart cities: Lessons from Seoul and San Francisco. *Technological Forecasting and Social Change, 89*, 80–99.

Porter, M. E. (1980): *Competitive Strategy*. Free press.

Teece, D., Peteraf, M., Leih, S. (2016): Dynmic Capabilities and organizational agility: Risk, uncertainty, and strategy in the innovation economy. *California Management Review, 58*(4), 13–35.

## Further Reading

Chappel, T. (1996). *The soul of a business: Managing for profit and the common good*. Bantam Books.

Connely, B. (2020). *Digital trust*. Bloomsberry.

Schiuma, G. (2011). *The value of arts for business*. Cambridge University Press.

Silberberg, P. (2016). *The ethical entrepreneur: Succeeding in business without selling your soul*. Morgan James Publishing.

Sutton, A. (2018). *People, management and organization*. Palgrave Macmillan.

### References on the WWW

Ben & Jerry's. www.benjerry.com/values (Last accessed: March 5, 2021).

Forbes. https://www.forbes.com/sites/cherylsnappconner/2013/03/04/the-8-great-challenges-every-business-faces-and-how-to-master-them-all/#5c6e89338914 (Last accessed: March 5, 2021).

Hiscox. https://www.hiscox.co.uk/business-blog/the-10-biggest-challenges-businesses-face-today-and-need-consultants-for/ (Last accessed: March 5, 2021).

Lean Methods Group. https://www.leanmethods.com/resources/articles/top-ten-problems-faced-business/ (Last accessed: March 5, 2021).

Patagonia. www.patagonia.com/core-values (Last accessed: March 5, 2021).

# How to Organize Knowledge Needed in Transformation?

Contents




I. T. Hawryszkiewycz, *Transforming Organizations in Disruptive Environments*,
https://doi.org/10.1007/978-981-16-1453-8_5

The earlier chapters stressed that good transformations must deliver value to people, businesses, and the environment. Designers must develop knowledge about the .complicated relationships. This chapter emphasizes the importance of knowledge. It is not only knowledge about what is happening now, but also knowledge of best ways to identify what is needed to improve systems and how to do it. It sees knowledge creation as continuous learning through identifying needs, suggesting solutions, and evaluating outcomes.

**Learning Objectives**

- Creating new knowledge
- Importance of knowledge in transformation
- Business processes for transformation
- Double diamond method.



## 5.1 Introduction

The previous chapters outlined the importance of values to people and organizations and ways to capture them. Now we look at what you do once you capture these values. Ultimately, the goal of any transformation is to deliver value to stakeholders. The question addressed in this chapter is how to convert the data collected, to knowleldge and how to üse the knowledge to transform a system to add to personal and organizational values.

To review earlier chapters, value is often associated with some monetary amount. On the other hand, value takes a more abstract view—like for a person it might be "My Well-being" or "My Health", for a city it might be "safety." For a business, it may be "being relevant and attracting customers". Ultimately you might reduce all these values to a monetary value but in early stages of design the more general term, value, is used. For a business, the value goal may be to improve the business brand; for a city, the goal is to become a smart city.

The goal of any transformation is to somehow realize the value of all stakeholders, businesses, and the environment. It is to create value of each stakeholder at the same time. Thus, a new transformation, not only provide value for stakeholders but should also add some value of a city and the environment. For example, a new building, should not reduce walkability by removing footpaths. It should, on the other hand, improve safety by making sure all footpaths are level. To design a transformation that satisfies stakeholder values requires the following:

- Knowledge of how to satisfy values and needs of people.
- Knowledge of how to satisfy the business values.
- Knowledge of how to add value to the environment and ways to respond to disruptions in the environment.

## 5.2 Developing Knowledge for Design

Knowledge does not, however, just appear. It must be developed, often gradually, from experiences and from data gathered from people. Such knowledge must be provided by people involved in the process with data that allow stakeholders' involvement. Data gathered on the spread of disease when analysed creates knowledge on ways to control its spread. Knowledge on hotel guest profiles will help hotel to plan menus, room service, and so on.

- *Data* are the basic reports captured as part of any activity.
- *Information* is how the data are organized to provide summaries.
- *Knowledge* is at a higher level—how do we use the information to identify issues and solutions.

## 5.2.1 Knowledge Creation as Continuous Learning

Knowledge creation is seen here as a process of continuous learning. For example, a sports team or player plays a game and use different tactics and observe on the outcome—did they win or lose? They analyse the outcome, discuss what better ways are needed to play better next time, agree to implement these ways, play the new way, discuss the outcome, repeating the cycle again and again. The cycle of continuous learning is widely applicable. It is happening today with COVID-19, where observations are made on the spread of the virus, alternate strategies discussed, some strategies are evaluated, outcomes assessed and used to develop further strategies. The learning cycle is applicable to any transformation.

Nonaka (1994) developed a well-known knowledge creation model for business transformation based on his experience in manufacturing in Japan. Nonaka's model is shown in ◘ Fig. 5.1. It has four quadrants. The four quadrants are parts of Nonaka's circular model of knowledge creation. They are called, socialization, externalization, combination, and internalization. These terms are sometimes seen as formal and in practice we use simpler terms: observations, proposing solutions, transforming systems, and evaluating solutions.

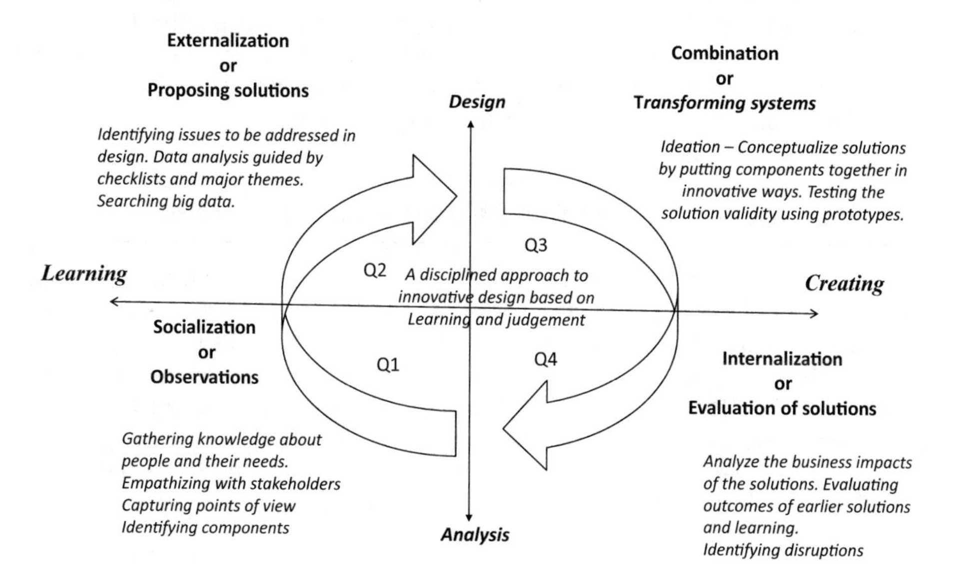


◘ **Fig. 5.1** Nonaka's knowledge creation model

What is important is that the four quadrants do not signify a process—they are ways that organizations learn through knowledge creation.

**Knowledge Creation in Summary**

Observations, (called socialization by Nonaka) is where people discuss what they are interested in, what they value and need, and generally share their tacit knowledge. Finding people values through discussion is an example of socialization.

Proposing solutions (called externalization) is to look at ways to use knowledge gathered in socialization to improve their system, define the issues to be addressed, and define what needs to be done, often what services need to be delivered.

Transforming systems (called combination) is where they integrate selected ways into the existing system—how to frame the services into the existing system. It includes knowledge about the components of the current system. These components or infrastructure is what can be changed in the transformation.

Evaluation of solutions (called internalization} is then the actual implementation of the services and monitoring whether the goals of the transformation have been achieved.



These quadrants are now described in more detail as follows.

### 5.2.2 Identifying Values—Socialization

In the first quadrant (Q1), we collect the most in-depth knowledge on the values of people and organizations. ▶ Chapter 3 described some ways to collect such data through observations, interviews, and other ways to gather what stakeholders value, where they see problems, or opportunities. They ask in-depth questions and make observations to identify what is happening, capture stakeholder points of view and their values, and any early suggestions and solutions that they have. What people do now, and what are their values, and where these values are not realized. In the case of the pandemic, we collect data through what is now called testing.

### 5.2.3 How the Knowledge Can Be Used—Externalization

In Quadrant Q2, we record why values are not being met. It is called externalization because here we discover about what is happening now, and how it affects our values. Here, designers identify major issues by discussion with stakeholders. These capture what people see as issues and disruptions, as well as opportunities in the current environment. In Q2, questions like "*What if* do something?" are asked to see how existing processes can be changed. Even when starting a design in a new organization, remember that any organization has any number of existing issues. These you often find from documents or from stakeholders—past and present.

### 5.2.4 How the Context Can Be Changed—Combination

In Quadrant Q3, designers develop ideas on how to solve the problems identified in Q2 and frame the solution into the system. Usually there are several ideas. In the case of a regular take-away, it may be possible for the restaurant to negotiate prices for special meals. Here brainstorming can be useful—get as many ideas as possible before you decide what to do. Solutions to implement these ideas are generated here; so is what needs to be done in the existing system to implement solution. Solutions can include new technologies, reassigning responsibilities, and training. In fashion design, solutions can include new ways of marketing and manufacture.

### 5.2.5 Continuous Monitoring—Internalization

In Quadrant Q4 is where we evaluate a transformation and then the cycle repeats. In that sense, the second quadrant defines requirements. These are implemented in Quadrant Q3 and outcomes evaluated in quadrant Q4. Then the process starts again, has the new implementation met values or have we learned something new from the current solution and can move forward to better solutions. It is here that we begin to identify best practices.

**▶ Example of Sporting Team**

In a game of sport

Socialization is where we describe the sporting rules and abilities of different players. We also describe what happened in the last game.

Externalization is where we might discuss the last game, why we lost or won, and what we can do to play better next time. "What if" questions are common here.

Combination is where we restructure our team and tactics for the next game.

We then play the next game.

Internalization is where we analyse how the new tactics worked in the current game, then go back to socialization.

Each cycle in ◻ Fig. 5.1 here is one game. ◀

Why is knowledge important?

Knowledge is important as any transformation must first find knowledge about the organization and its stakeholders. It must also find and develop knowledge about ways to add value by finding out what new ways and methods are needed to do so. Then knowledge is needed on ways to introduce these new methods and find out if they have added to stakeholder value.

## 5.3 What Are Practical Ways to Develop Knowledge?

The question now is how to put the knowledge creation cycle into practice. ◘ Figure 5.2 shows what needs to be done. Ultimately, knowledge is created through a transformational process like that shown in ◘ Fig. 5.2. Such a process is itself a business process with a goal and outcome.

**Organizing a Transformation**

Recoding knowledge collected and created.

Activities through collaboration to use collected data and information to create new knowledge.

Organizing the transformation into a business process.



Initially we find data through interviews and documents like those discussed in ▶ Chaps. 3 and 4. Post-it notes are often used in quadrant Q1 and often quadrant Q2. Here stakeholders might stand in front of a board and quickly post their stories, suggestions, and ideas. More formal models are developed in quadrants Q3 and Q4 that outline ideas for solutions, and the rationale used to adopt them.

The activities used in analysing the knowledge use design-thinking methods and are described next. Organizing the transformation process is described in ▶ Sect. 5.5.

### 5.3.1 Reasoning to Create Ideas and Solutions

Designers need to find reasons of why a solution meets stakeholder values. Complexity presents challenges to transformation because of complex relationships that must be simplified make such reasoning difficult. In the past, most problems were well-defined

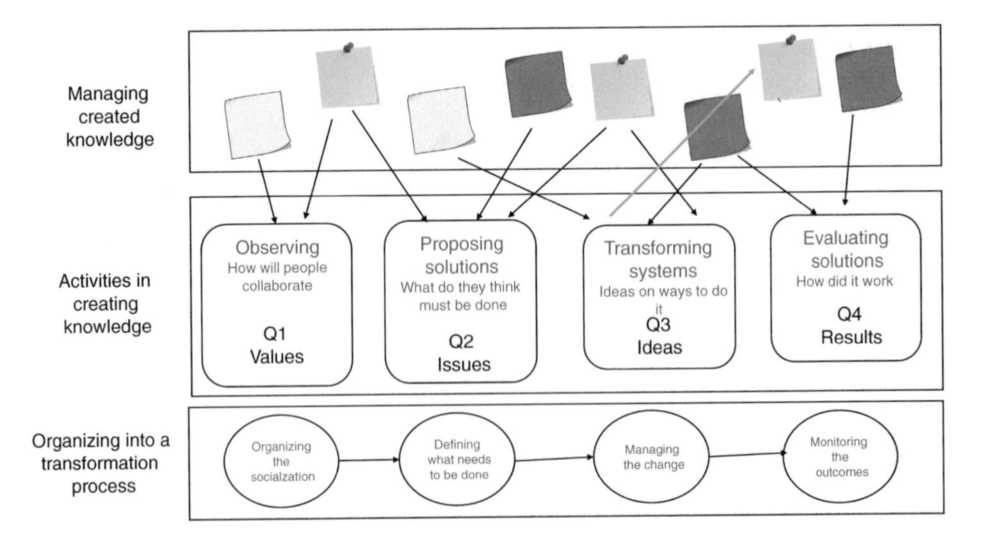


◘ **Fig. 5.2** Managing knowledge creation

and could be solved in a systematic and often analytical and systematic way. The knowledge was explicit and often well documented. In such design, we know where we are and where we want to be and how to get there following well-defined steps using deductive reasoning.

### 5.3.2 Using Deductive Reasoning

Suppose we are designing a building. Given the proposed height of the building we can compute or deduce the steel beams we need to ensure it stands. This calculation can be done precisely by using the mathematics that form part of building science. We can exactly define what we want and then reason using deductive method of what we need to do to satisfy our goal of making sure the building stands.

But human values must also be considered in building design. Decisions on building height often have a social focus. For example, who decides the height of the building; is it the building owner or the municipality. The internal layouts in the building begin to focus on the values of occupants on how they work and live.

But what if no rules exist at all? If we know exactly what each part does, then we can use rules to deduce what a system composed of such parts will do. In fashion design, the emphasis is on values, especially in the trends found in what people like to wear. Such trends cannot be found exactly using mathematical or some other precise form of reasoning. Abductive reasoning is often used here.

### 5.3.3 Using Abductive Reasoning

The best way to describe abductive reasoning is by suggesting an idea and see it works. The emphasis is more on brainstorming and asking "what if" questions. Thus, faced with an issue, we ask what if we do some specific thing to address the issue.

In the fashion industry, the design is less formal and emphasizes people values and trends that cannot be mathematically defined. However, we also need to consider technical factors, like choosing textile materials or mass production to reduce the cost. Both cases illustrate the greater emphasis on social values and add to design complexity.

Abductive reasoning is increasingly relevant in wicked environments.

To identify what is happening and discovering the detailed nature of the problem, especially where people's values are not being met. Such in-depth questioning discovers relationships and issues that may not be obvious initially: but how they impact on people's behaviour may not be initially obvious. It is not possible to have precise deterministic measures to measure the outcomes. Instead, there is more emphasis on qualitative rather than quantitative outcomes—delivering value to stakeholders. For example, usually restaurant meals are judged by their taste rather than calories—how to measure taste. Qualitative values are often used here—is tasting sweet, better than tasting yummy? Then we need to provide an acceptable solution that meets these values, at least partially.

Perhaps one of the most recent examples of the importance of knowledge is that of the 2020 COVID-19 pandemic. The goal has been to eradicate, or at least, control the spread of the coronavirus. Data had to be gathered on its spread, ways to minimize the spread developed, suggest actions, and monitor the outcomes of the actions. Knowledge is critical here—knowledge about how the virus spreads, knowledge of how to stop the spread, and knowledge of what is the best way to manage in a disruptive environment.

## 5.4 Design Thinking in Knowledge Development

The next question is how we gather and create knowledge. This book uses design thinking as one way to develop the knowledge. Buhl and others (2019), for example, describe the suitability of design thinking in complex systems as it supports all the quadrants in ◘ Fig. 5.1. Design thinking is a human-centred process where design focuses on people and their values. It provides the visualizations, and the in-depth questions seek out the knowledge. Here teams use these methods to jointly brainstorm to identify issues of concern and ways to address them. Solutions are often multi-disciplinary as they must satisfy many stakeholders. The solution focus is on teamwork and brainstorming to use stakeholder tacit knowledge.

- Identifies stakeholder values and needs
- Develops solution ideas through brainstorming
- Focuses on visualization and models to encourage creativity and innovation
- Supports experimentation through prototyping

Perhaps the most significant characteristic of design thinking is that solution is approached gradually, as outlined in the previous chapter. It is like going around the circle in ◘ Fig. 5.2 several times, often continually. Each time part of the situation is improved, we at the same time discover new problems and issues for the next transformation step. Transformation is thus often incremental in wicked environments—solutions emerge rather than being predefined.

Design thinking emphasized diagrammatic and visual tools whose goal is to stimulate creativity. Some of these are described in the supplement at the end of this book.

Another important characteristic is its emphasis on problem finding. It has often been said that it is important that we focus on the right problem. In the words of Albert Einstein "*If I had an hour to solve a problem and my life depended on the solution, I would spend the first 55 minutes determining the proper question to ask … for once I know the proper question, I could solve the problem in less than five minutes.*"

An important activity in design thinking is on questions, continually questioning what is going on.

- In Q1—*What* are stakeholder values? *What* is happening?
- In Q2—*Why* is something not meeting values?
- In Q3—*What if* we do something?
- In Q4—*How* will we do what is proposed? Has it worked?

Using design thinking requires a more in-depth analysis of what is happening. For example, when do people want access to vegetarian food, is it available closely, but only for limited hours. Are they within short distance, but transport closes too early? Do you need selection or standard?

### 5.4.1 Example—Knowledge in the COVID-19 Pandemic

As an example, ▫ Fig. 5.3 uses the guidelines provided in ▫ Fig. 5.1 to show the complexity of knowledge found in most transformations. It shows the knowledge developed in managing in the pandemic.

▫ Figure 5.3 shows only some parts of the story. But it illustrates:

- The complexity because of the number of relationships as shown by the arrows.
- The circular nature as an example shown by the thick dotted arrows.
- The number of stakeholders and their activities, where in a complete model would be many more than shown here.

Diagrams like ▫ Fig. 5.3 are used later in this book to illustrate the complexity of transformation and as a reminder of what needs to be done. They show the overall picture of what knowledge is needed. Ways to find and create such knowledge are described in the following chapters. In diagrams like ▫ Fig. 5.3, the knowledge is inside the circle, whereas methods used to develop the knowledge are outside.

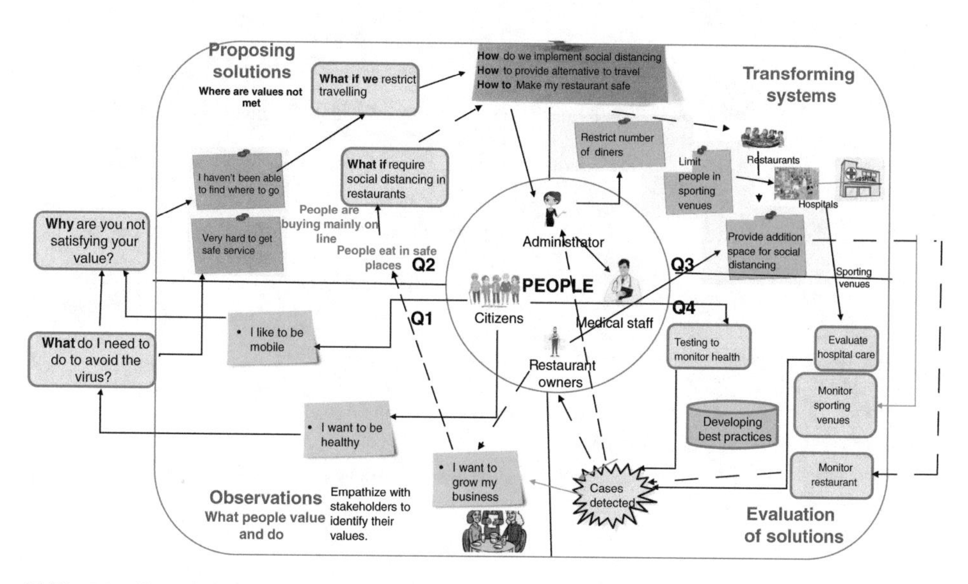


▫ **Fig. 5.3** Knowledge development in the pandemic

One of the important things to notice is the *circular nature of knowledge development*. ◘ Figure 5.3 shows that many citizens value health as important, whereas businesses like restaurants want to grow their activities. As a result, people want to eat in safe places, administrators want to preserve social distancing, health administrators want to contain the virus. ◘ Figure 5.3 shows the circular nature of one part of the complex system—restaurants. Only one such circle is shown (by the heavily dotted lines). The circular nature is one way to deal with complexity—there would be many such circles in any system, and businesses need to see where they fit into such a circle to see what actions they need to take.



A restaurant owner, who wants to grow their business, finds that people are hesitant in going to restaurants (Q2). The question then is HOW to convince people that going to restaurants is safe. One solution is to maintain social distancing. Social distancing may be implemented by rearranging seating but may not result in an increase of diners. If the number of diners does not increase, the restaurant owner may then look for another alternative, such as, for example, beginning to offer take-away meals. ◘ Figure 5.3 also shows how information gathered during transformation is recorded on post-it notes and placed in the most relevant quadrant. It also shows the complexity of dealing with the pandemic by the number of relationships, and circles—in fact, there are many more than shown in ◘ Fig. 5.3, showing the complexity of dealing with the problem.

◘ Figure 5.3 also summarizes some of the kinds of data collected and sorted into the quadrants.

| Quadrant | Knowledge needed |
|---|---|
| Q1—what are the stakeholder values? | People's values such as health. Commerce, well-being, and their general ability to be mobile<br>These were described earlier in ► Chaps. 3 and 4 |
| Q2—How do current events (such as the pandemic) effect values? | How to reduce the negative impact of the virus on my health?<br>How to maintain the mobility to reduce impact on commerce? |
| Q3—How to stop adverse events? | Ways to avoid catching the virus<br>What if we have social distancing?<br>What if we have lockdowns?<br>Define lockdowns |
| Q4—What are the best ways to improve values? | Monitor what works<br>Monitor the adherence to rules<br>Testing of the community<br>Identify disruptions that call for action<br>Identify best practices as progress is monitored |

### 5.4.2 A Commercial Example

Let us go back to our hotel, which was identified to be composed of several systems.

The kinds of questions we ask are summarized below.

- *Step 1* in Q1—Begin with values by asking questions like "*What* kind of guests come to your hotel?"—the answer shown by the blue post-it.
- *Step 2* in Q1—*what* is happening now?
  How long do your guests stay?
  What services do they use?
- *Step 3* in Q2—try to find the reason for guests not returning.
  *Why* are there no restaurants?
- *Step 4* in Q3—what can we do to address the gap?
  The questions here take the form *what if* we do something?
- *Step 5* in Q3—The next question *how* will we do something? Will it work?
- *Step 6* in Q4—Then we ask, has it worked and improved.

The sequence of questions—*What, Why, What if, and How* is a typical set used in design thinking. It can be expressed in different ways. Liedtka (2013) has the sequence, *What is*, *What if*, *What wows*, and *What works*. These correspond closely to those above. One difference is in what wows and excites stakeholders rather than what if to stress that any proposed solutions are something worth doing.

The important thing to remember is that we go around the circle many times as design never stops, but is a process of continuous improvement. Another technique in the diagram is to show any comments as post-it notes—a technique quite common in design thinking.

### 5.4.3 Applying to Business

Knowledge development for a hotel, as shown in ◘ Fig. 5.4 using the circle diagram, is there for illustrative purposes. The amount of information designers need could not be fitted onto the circle.

Again, it illustrates several segments of hotel guests—backpackers. There are also the hotel employees that need to be considered in any design. Such large numbers lead to a greater variety of values and needs. ◘ Figure 5.4 only shows the number of different guests. It shows the following:

- Has many kinds of guests, including, tourists, backpackers, families, and business travellers.
- These guests have different needs and expect the hotel to provide services for each.
- There is a wide variety of services needed that include providing breakfast, relaxation for elderly guests, and so on.

There are also issues like irregular working hours, introducing new technologies as guests become more technology literate, improving maintenance systems, and updating of guestrooms all adding to the complexity. The hotel vision is to be able to respond quickly to customer requests.

◘ Figure 5.4 also shows one circle—meeting the needs of business guest by providing technologies to enable them to carry on with their business.

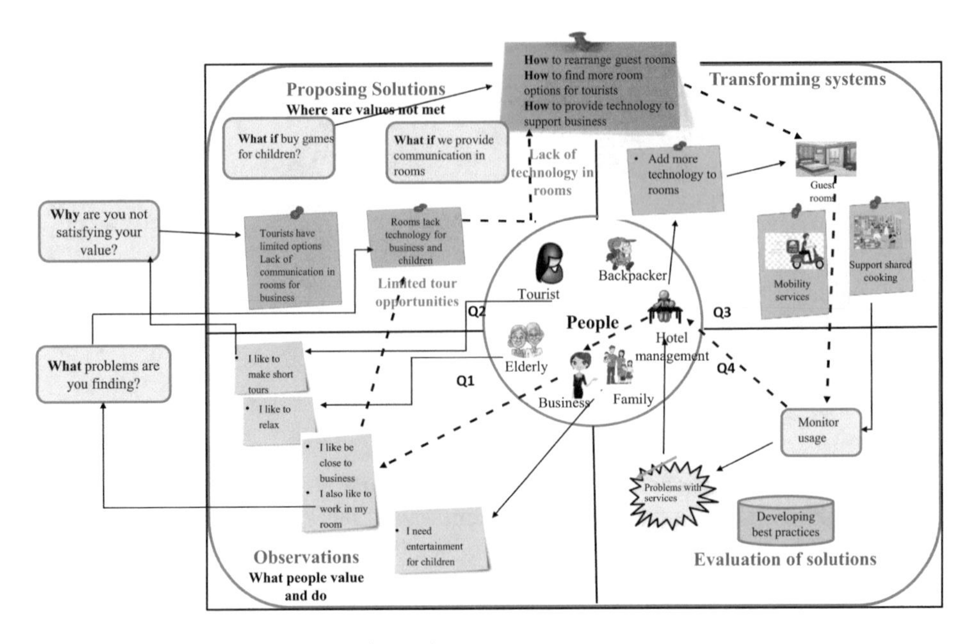


Fig. 5.4 Knowledge development in hotel management

There are also sets of circular activities like that shown for satisfying business needs through technology support in guest rooms. There are many more, again demonstrating the complexity of any transformation.

Figure 5.4 is still a simple example of an organization, but it illustrates some ideas that are now found in practice. The first is that we cannot make complex systems simple. In that case, the hotel in Fig. 5.4 would simply provide one kind of room to cater for one type of guest—catering for one type of guest would reduce the number of guests, its revenue and, ultimately its survival.

## 5.5 Integrating Transformation into Organizational Processes

Management eventually wants to see some outcome of design. They organize design into a formal process that can be monitored to ensure there is progress to a desired outcome.

### 5.5.1 British Design Council—the Double Diamond Method

The double diamond method, shown in Fig. 5.5, is an increasingly popular method that is proposed as a standard by the British Design Council (see reference).

The phases of double diamond.

The double diamond method closely corresponds to the way that information is developed, and decisions are made in wicked environments described in Fig. 5.2. Briefly:

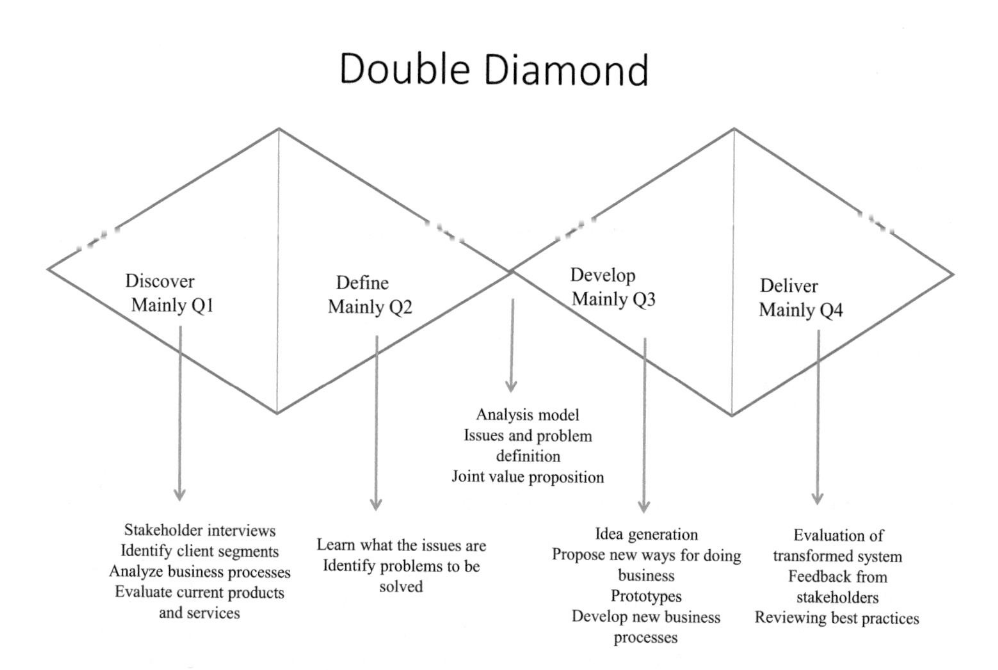


**Fig. 5.5** The double diamond method

- Begin with discovering what is happening in the system and the stakeholder values. Define client segments. Analyse client experience with products. These activities closely correspond to quadrant 1 of Fig. 5.2.
- Make sense of identified issues and define problem. The correspondence here is to Quadrant 2.
- Develop ideas and ways to satisfy stakeholders as in quadrant 5. Check with suitable interfaces and scenarios. Analyse costs and benefits.
- Evaluate solution outcomes—Quadrant 4.

This book closely follows the Double Diamond method in principle. ▶ Chapters 3 and 4 describe the values that form the basis of the decisions. ▶ Chapters 5 and 6 then describe the first part of the diamond resulting in a design.

### 5.5.2 Stanford dSchool

The Stanford dSchool uses a design thinking process shown as follows to develop creative solutions. Although design thinking has earlier origins, the way of using design thinking, as shown in Fig. 5.6, originated from ideas from Hasso Plattner et al. (2011), the former Chairman of SAP. It also led to a design process that starts with empathizing with stakeholders to identify their values and lead to innovations (in product or service) acceptable to stakeholders.

It is important to realize that creativity and new ideas are not only needed at one part of the process. It takes place throughout the design, starting with identifying needs and then identifying ways to satisfy these needs in innovative ways. Prototyping is often used for stakeholders to jointly develop solutions with designers.

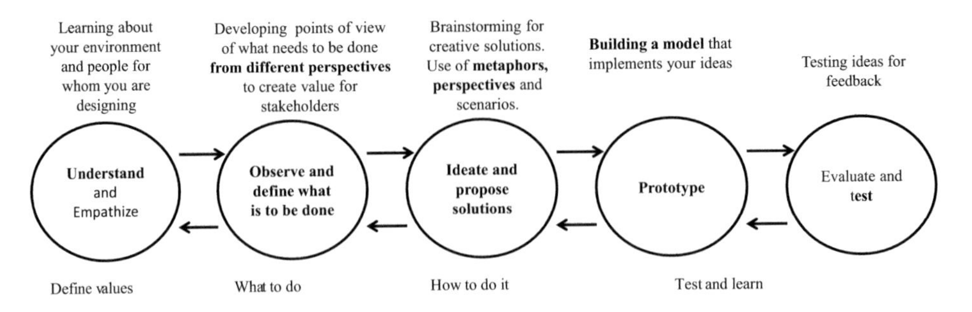


**Fig. 5.6** Fostering creativity in the Stanford process



**Summary**

In summary, this chapter introduced design in complex environments. It set the scene for the rest of this book. It defined the principles of design and illustrated with simple examples. It first described the need to collect knowledge about the design environment and ways of processing it to get a good outcome. Good designs not only focus on personal values. They must also include organizational values. It described a generic process (Fig. 5.2) as one that is used in the following chapters to help one to generate knowledge to make the best choice in their system.

It also identified some guiding principles in managing complexity in a value-based approach.

- One is to find what is going on in the system through rigorous questioning of what is happening. It identified one stream of questions, as what, why, what if, and how.
- Then we identify the services needed to satisfy these needs.
- We then look at how we can frame the service into the infrastructure—and whether the infrastructure needs to be improved.

The important takeaways are as follows:

- Alignment of personal, business, and city values.
- Abductive reasoning or developing solutions by almost trial and error.
- Framing the solutions into an existing context.

## Exercises

Define knowledge needs for the earlier case studies in the four quadrants.

### Restaurants and Hospitality Responses from the COVID-19 Pandemic

Many restaurants (and hospitality) providers are being adversely affected by the disruption caused by the pandemic.

Identify a restaurant you might be familiar with. Identify its systems such as serving customers, preparing meals, ordering produce. Identify where COVID-19 poses threats and what is happening to reduce the threats.

**Food Supply Is One of the Most Critical Goals in Any Society**

Food supply chains include many businesses—farmers, processors, supermarkets, among others.

Describe the knowledge needed by these businesses and the disruptions possible in each business and as products move between businesses.

### Vegetarian Diners

Restaurants need to grow customers to survive. To grow customers, they need to develop knowledge about diners' preferences and way to satisfy diners' preferences and satisfy them. ◘ Figure 5.7 illustrates the knowledge development process that focuses on one stakeholder – a vegetarian.

Develop the kind of questions you would need to develop knowledge on the best way to provide services to vegetarians. Start with observations of what vegetarians' value. Some observations are shown in the bottom left-hand corner, as for example, "I like good vegetarian food". Go into more depth to identify kinds of dishes. We then look at what is happening now, what are stakeholders doing and where do they find that their values are not being met.

In your team you might try using the knowledge circle to post post-it notes in the way shown in ◘ Fig. 5.7.

However, more in-depth questioning can lead to more information. For example, stakeholders here may not wish to travel any distance to a vegetarian restaurant, or their need is for specialized vegetarian food. They may all work in the same or close buildings. You also find that there are close by restaurants that close early in the evening. Most of the stakeholders work late and want vegetarian food late at night.

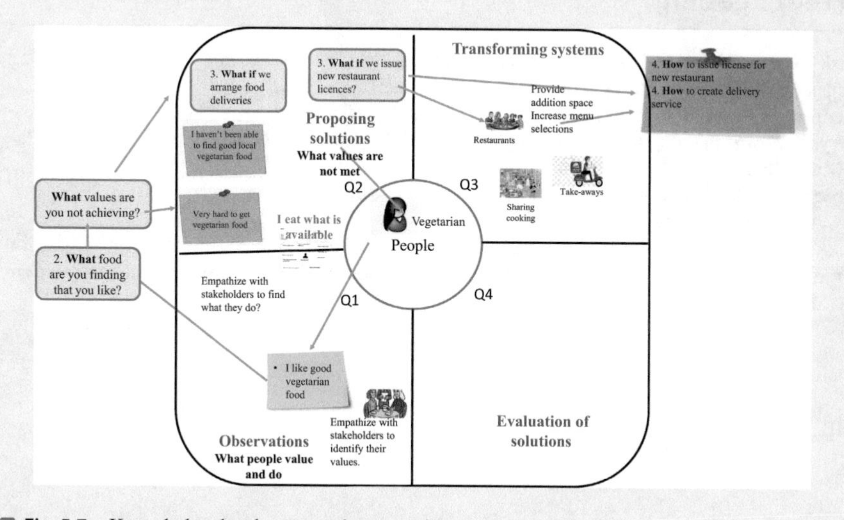


◘ **Fig. 5.7** Knowledge development about serving vegetarian diners

- *Questions* in Q1—Begin with values by asking question like "*What* are your preferences in food?
- *Questions* in Q1—*What* is happening now?
  - We eat now what is available. Do you like it? Why not?
  - Why? Because there are no nearby vegetarian restaurants.
- *Questions* in Q2—Try to find the reason for values not being met.
  - *Why* are there no restaurants that offer vegetarian dishes?
  - What are we doing to find vegetarian food?
- *Questions* in Q3—What can we do to address the gap?
  - The questions here take the form *what if* we do something?
- Questions in Q3—*How* will we do this? Will it work?
- Questions in Q4—Did we improve by changing what we do?



Now begin to identify any circular activities.

## References


Buhl, A., Schmidt-Keilich, M., Muster, V., Blazejewski, S., Schrader, U., Harrah, C., Schaefer, M., & Sussbauer, E. (2019). Design thinking for sustainability: Why and how design thinking can foster sustainability innovation development. *Journal of Cleaner Production, 231*, 1248–1257.

Liedtka, J., King, A. Bennett, K. (2013). Solving Problems with Design Thinking. Columbia Business School.

Nonaka, I. (1994). A dynamic theory of organizational knowledge creation. *Organization Science, 5*(1), 14–37.

Plattner, H., Mienil, C., & Leifer, L. (Eds.). (2011). *Design thinking understand – Improve – Apply*. Springer-Verlag.


## Further Reading


British Design Council. *A study of the design process*. http://www.designcouncil.org.uk/sites/default/files/asset/document/ElevenLessons_Design_Council%20(2).pdf

Davenport, T. H. (2005). *Thinking for a living*. Harvard Business School Press.

Hawryszkiewycz, I. T. (2010). *Knowledge management: Organizing knowledge based enterprises*. Palgrave Macmillan.

Holbeche, L. (2018). *The Agile organization*. Kogan Page Stylus.

Innovate UK. (2015). *Design in innovation Strategy 2015–2019*.

Nonaka, I., Kodama, M., Hirose, A., & Kohlbacher, F. (2014). Dynamic fractal organizations for promoting knowledge-based transformation – A new paradigm for organizational theory. *European Management Journal, 32*, 137–146.

# Identifying and Solving Problems

Contents

# Describing the System

Contents




I. T. Hawryszkiewycz, *Transforming Organizations in Disruptive Environments*,
https://doi.org/10.1007/978-981-16-1453-8_6

This chapter begins to describe how transformations create value for businesses and their stakeholders. The first step is to create models of knowledge so far captured about organizations and their business. These models are then used to discuss potential changes and transformations that lead to value improvements.

**Learning Objectives**

- Developing system models
- Identifying data needs
- Developing journey maps
- Combining into model
- Importance of traceability



## 6.1 Introduction

▶ Chapters 3 and 4 described the values of people, businesses, and cities. ▶ Chapter 5 then described the knowledge that is needed and created in a transformation. To satisfy both personal and organizational values require detailed knowledge of how the organization works. The trend is to build models to help designers identify how organizations work and how to transform them to meet stakeholder values. The models are then used to see what detailed changes can be made in the transformation.

The models are often diagrams or even sketches; they are not mathematical, but show the connections between different parts of the organization. These models are used later to identify potential changes to the ways these parts work together, and to add new parts. A model will show all the components, all stakeholders, and how they interact to create value. A model of the organization at this early design stage is called an analysis model.

The analysis model shows what is happening now. It is divided into:

- The *systems part* which shows the stakeholders, their workplaces.
- The *data part* which shows the data and information in the system.
- The *process part* which shows how the stakeholders carry out their activities.

The analysis model also includes the stakeholder responsibilities and the relationships between units and the stakeholders. Models are often seen as repositories of knowledge about a system. One way to see building a model during analysis is as collecting knowledge about a system.

### 6.1.1 Models as a Repository of Knowledge

As a revision, Nonaka's circular model can be used to describe what knowledge is needed. We use the example of a shopping centre or mall. ◘ Figure 6.1 shows the knowledge circle, which was defined earlier in ▶ Fig. 5.2. It shows the major stakeholders in the retail side of shopping centre activities—the majority of which are shop owners, customers, and centre management. There are of course others, like

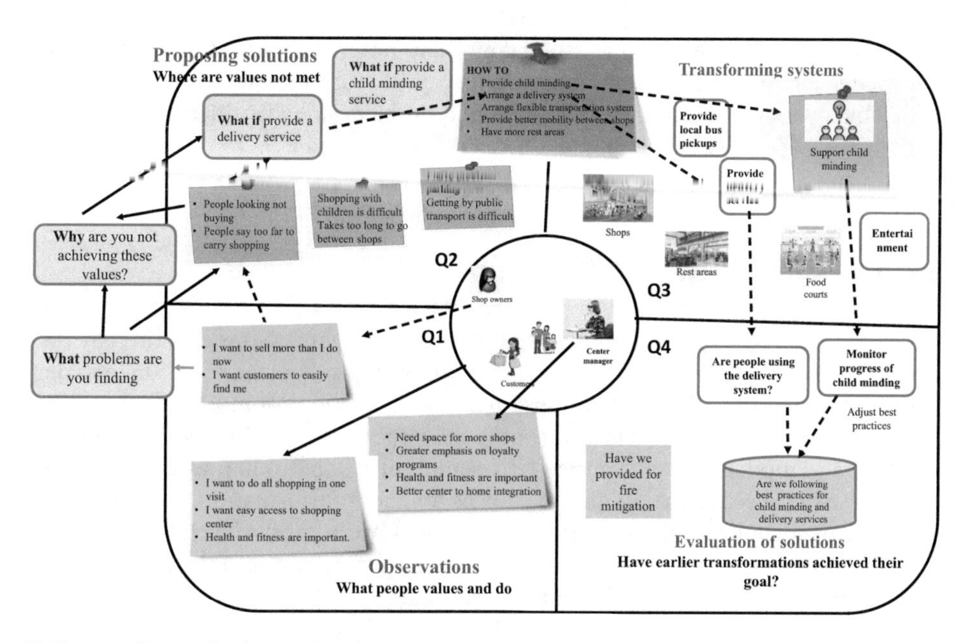


**Fig. 6.1** Developing knowledge about a shopping centre

maintenance people, suppliers, and operators of various facilities. Figure 6.1 again illustrates the complexity of satisfying values of all stakeholders. It focuses on one cycle—that of increasing sales outcomes of shop owners.

From a design thinking perspective, it is common to gather stories of what is happening and why it is happening, often by collecting stories. The methods used to collect these stories are interviews, reading about the industry, and stories or narratives of how the business runs. The captured stories are shown as post-it notes, which are frequently used in industry. The stories add knowledge to all quadrants of knowledge creation.

Q1. Find who the stakeholders are, their values and needs. Figure 6.1, for example, shows stories gathered from shoppers, shop owners, and others. It describes what people do, and what the concerns are. Shopkeepers, for example, want to increase their sales.

Q2. Identify stakeholders' issues and problems of meeting values by asking WHY questions and propose HOW to deliver solutions. One issue faced by shopkeepers is that their customers find it hard to carry their purchases to their cars.

Q3. Propose ways to transform the organization to resolve problems and add value by providing a delivery service, as well as a child-minding service so that parents have more time to shop.

Q4. Gather knowledge by monitoring whether and how the transformation that provides a delivery service has added to value, and to suggest further improvements.

One point to remember is that businesses in the same industry often work differently. Each shopping centre is different in the way they arrange shops, the kind of products they sell, and the entertainment they provide. Figure 6.1 shows stories collected as post-it notes in Q1 linked to the different stakeholders. These are of course only illustrative and, in practice, there will be many more notes. You can if you like draw a knowledge circle like ◘ Fig. 6.1 and use it to post stories about a shopping centre with which you are familiar. In Q2, issues and problems are identified and described later in ▶ Chap. 7. Then ▶ Chap. 8 describes what to do in Q3 to provide solutions to problems identified in Q2.

## 6.2 The System Part

Visual models are now increasingly used to describe organizations and the way they work. Persona maps described in ▶ Chap. 4 showed a visualization of stakeholders and their values. This chapter develops further visualizations or models to show what stakeholders do in the organization and the processes they follow. Modelling is important as models are an accurate representation of what is happening now. They can be used in discussions on design choices for the transformation. The modelling methods or visualizations in the analysis model are almost sketches that enable designers to visually see the relationships in the context where design takes place, and to describe and discuss potential transformations.

### 6.2.1 Rich Picture Showing How Everything Fits Together

Rich pictures show the business units, stakeholders, and the activities of people in the organization. The rich picture can also show what the systems are and where the knowledge can be found. The rich picture shown in ◘ Fig. 6.2. illustrates the infrastructure relevant to shopping activities. The goal is to later encourage ideas that lead to innovative designs by seeing the whole organization and how changes to its parts or connection can add to value.

Often, to draw a rich picture like that shown in ◘ Fig. 6.2, it is best to ask questions such as those below and add to the rich picture based on the answers.

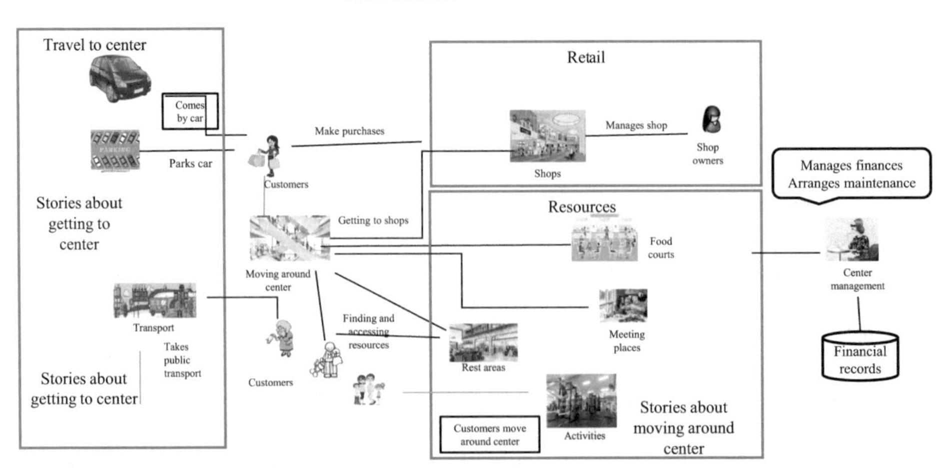


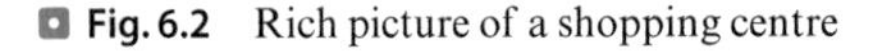
◘ Fig. 6.2 Rich picture of a shopping centre

- Who are the stakeholders? Look at the stakeholders identified earlier and create an icon for each of them. ◘ Figure 6.2 shows, for example, icons for customers and shop owners.
- Where do they work? Draw the place of work: for example, how a shop is organized.
- Who do they interact with? Shows the relationships between the various entities by linking them. Links should also be labelled with the data passing across the link.
- What do they do? For example, customers visit shops, park cars. Purchase goods and engage in other activities.
- Show parts of the shopping centre infrastructure such as meeting places and others.
- What data do they need?
- How do they communicate?
- Group the objects into systems. The rectangular boxes group the activities into three systems.

As analysts ask questions, they often record the answers on the rich picture like that shown in ◘ Fig. 6.2.

The system model or rich picture shown in ◘ Fig. 6.2 is for illustrative purposes and is by no means complete. It shows the following:

- The use of meaningful icons to make the diagram easier to understand.
- Lines showing relations between stakeholders as, for example, customers buying in shops.
- Data useful to stakeholders as, for example, financial data for the centre manager.
- Major responsibilities like financial management for the centre manager.

Systems diagrams like ◘ Fig. 6.2 also provide guidelines for design. Often designers transform systems in parts. Thus, for example, one person can look at travel to the centre, another as ways to facilitate shopping, and still another at activities in the centre. They of course have to make sure that their designs fit together.

Designers often have a system diagram for the whole system. Then each of the major parts can be expanded into a more detailed diagram. For example, the retail part of ◘ Fig. 6.3 can be expanded into a more detailed form showing, for example, how shop owners interact with their customers or suppliers of items for them to sell. In the case of food courts, models can be used to find ways for food courts to be COVID-19 safe.

In your model, you might initially highlight where there are issues identified by stakeholders.

## 6.3 Data Part

Data is part of any system. It is needed to support virtually all activities in any system.

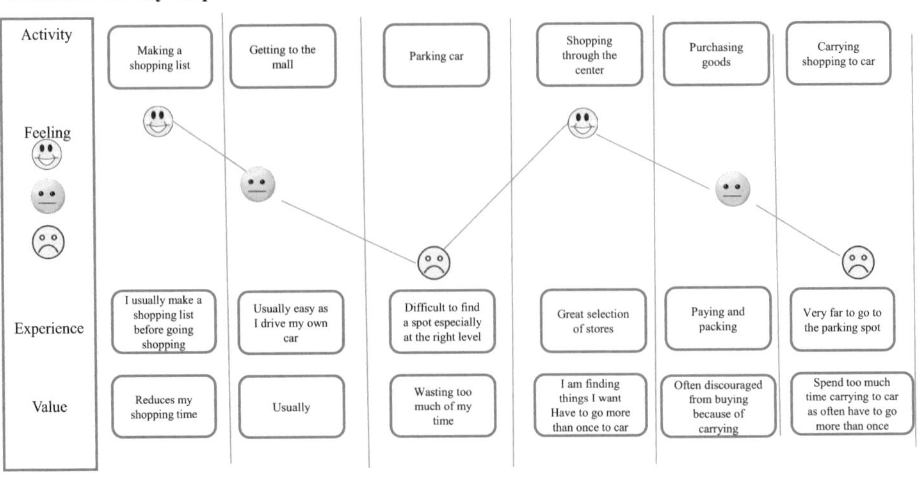

Fig. 6.3 A journey map

Data includes the following:

- Planning data.
- Management data defines the resources in the system and their current status.
- Operational data shows the what is happening at any point in time. For example, what are today's sales, or what is the status of a selected delivery.

Data can be described as a set of data and eventually as detailed items that make up databases. The data includes the following:

- Details of the infrastructure, as for example, the number of shops and what they sell.
- Amount of activity, as for example, people coming to the centre and what they do.

Later, design considers where the data is needed, how it can be found, and what is important, how it will be used to lead to better solutions.

## 6.4 Process Part

The next dimension shows how people work and the processes they follow. The most popular tool used here are journey maps.

### 6.4.1 Journey Maps Describe Processes Followed by Stakeholders

Journey maps describe how people carry out their work in the design context of the shopping centre. For example, Fig. 6.3 shows a journey map for a customer making a purchase. It shows the steps followed by a shopper to travel to the shopping centre, make purchases, and get the shopping home. Each step is seen as a touchpoint as it shows how stakeholders interact with each other and with parts of the system.

Touchpoints in ◻ Fig. 6.3 show a customer's experiences when they purchase something in the shopping centre. It shows how they feel at each step, what data and information they need, and whether they meet their values.

- How does a journey start—what is the first step?
- What is the next and the following steps?
- At each step—record stakeholders' feelings.
- At each step—record what value they have achieved.
- At the end, summarize the pains and gains.

Then each touchpoint itself can be described in detail. Such description should be provided at each touchpoint to describe the following:

- What the stakeholder needs to do at that touchpoint.
- The data needed to raise awareness of what is possible for them to do.
- What data is needed to do what they want.
- What option is most likely to be chosen.
- What data is needed to select an option.

Record what you find out in tables like that shown in ◻ Table 6.1.

**Journey Maps and Data**

Journey maps and their associated touchpoints are important as they identify the data that stakeholders currently have. It is important here to identify what operational data is needed by stakeholders in a process. Later, in design, we find ways to make operational data available to create what are increasingly seen as smart businesses—business that specifically fits into their environment and how they use data to become relevant and deliver timely outcomes to their customers.

Journey maps and the associated touchpoints are one of the most common tools used in design thinking, and consequently are of great value in organizational transformation. It is here where data is captured and created. It is also here where designers align personal values and organizational.

Journey maps come in different formats, and readers may wish to go to Google and look at the variety. It is, however, important to remember that journey maps are the processes followed by people in any organization. Process design is where there has been great emphasis in the current COVID-19 pandemic, how to adapt and simplify existing processes to enable remote work, or to maintain social distancing. It is important now to look at each step, what data it needs, the technology needed to support collaboration at the step.

Table 6.1 Touchpoint description

**Journey map: Purchasing of goods**

| Touchpoint: Customer purchases items | Customer | Shop assistant | What are data needs |
|---|---|---|---|
| Awareness (finding what is possible) | What items can I find that I like? How do I get them home? | Provides advice to customer | Finding something the customer really likes |
| Options (ways to evaluate options) | Select item to purchase. Make payment | Finding the best way to deliver a sale | Location of customer Items to be delivered |
| Research (on ways to evaluate options) | How to carry purchase to transport | Advise on delivery options | Estimate of delivery time |
| Selection and action | Rank various options | Data on item comparison | |
| Services used by person | Discussion with shop management | | |



### 6.4.2 Some Practical Suggestions for Drawing Journey Maps

There is no such thing as a standard way to draw journey maps. Journey maps are generally adapted to the situation. They, however, have important common goals. Most journey maps focus on customers and are used to find the best ways to interact and keep customers. The journey map also includes measures at each touchpoint—in Fig. 6.3, their feeling, experience, and value that stakeholders, in this case a customer, see in the step. For example, in Fig. 6.3, customers are happy with the selection of what they can buy, but do not have to carry purchases to the car.

How do you draw journey maps? Again, brainstorming and post-it notes can be useful here. Figure 6.4 is an example. Journey maps basically start from the question "What do stakeholders do?" Thus, for example customers go shopping. Then the question may be how they start. The first step is to go to the mall. Then put up a post-it note "Going to the mall." What do they do next? They park the car—put the next post-it notes.

### 6.4.3 Extending Journey Maps

There may of course be many journey maps in a system. We showed a customer journey map. There is also a journey map that shows how shop managers select and bring in products to sell. There are also journey maps on how keep the centre clean and safe, which will include centre management.

It is important to have persona maps of both customers and managers. It is from the managers that we discover organizational values; then we look at customer values

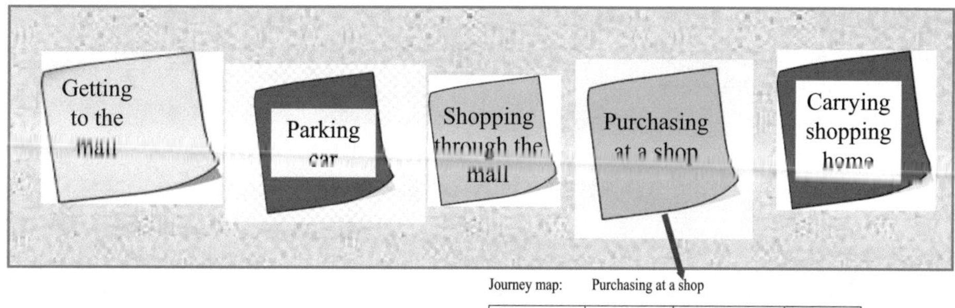


Journey map: Purchasing at a shop

| Touch point: Customer purchases items | Customer | Shop assistant | How is value created |
|---|---|---|---|
| Awareness (finding what is possible) | What items can I find that I like? How do I get them home? | Provides advice to customer | Finding something the customer really likes |
| Options (making a choice) | Select item to purchase. Make payment | Wraps purchases and processes payment | |
| Research (what is the best option) | How to carry to transport | Advise of delivery options. | Minimizing effort at carrying |
| Selection and action | Rank various options | Data on item comparison. | |
| Services used by person | Discussion with shop owner | | |

**Fig. 6.4** Developing a journey map

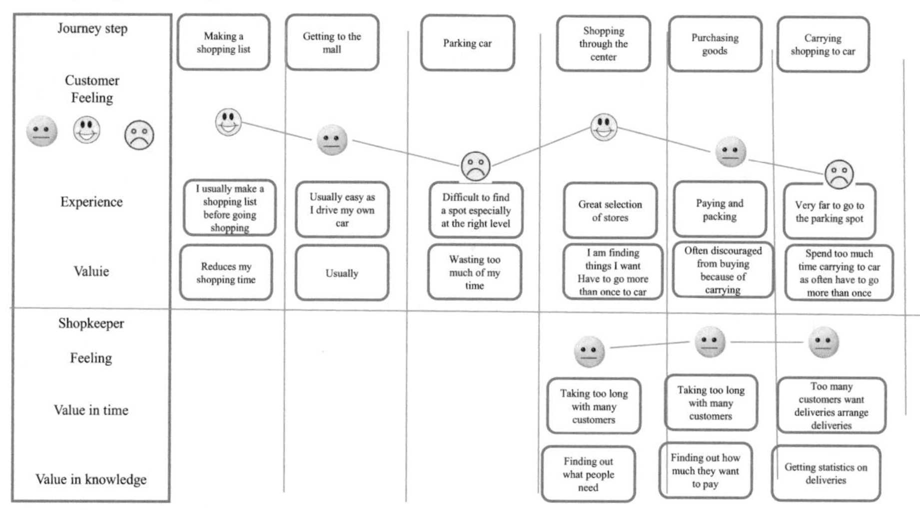


**Fig. 6.5** Journey map with more than one stakeholder

and see how they match. It is also important to have a number (preferably an exhaustive list) of customers. This list can be mothers with children, the elderly, or visitors browsing through the centre, as well as others. Figure 6.5 shows one such persona map for customers.

There are also different reasons to use journey maps. So far, we showed journey maps for a single stakeholder, customer for Fig. 6.3. Many journey maps, however, involve more than one stakeholder. The journey map in Fig. 6.5 is a case where another stakeholder is involved for part of the customer journey. The other stakeholder is the shopkeeper. The feelings of the shopkeeper and their value from the journey may be different.

## 6.5 Combining into an Analysis Model



Any business, city, or organization incudes people, business units, processes, and the relationships between them. A model of the whole system must include visualizations or model components for all of these. Furthermore, they must fit together to show the whole picture of what is going on. How to fit everything together is important in any proposed transformation because a change to one part often impacts the other parts. Such impacts must be predicted and included in any transformation. The relationships between the different parts of the system model can be shown in two ways—by links drawn between the different parts, and by naming consistency. The links can be shown visually, as shown in ◘ Fig. 6.6. Naming consistency is equally important; visualizations describe the same system and must use the same name for the same thing or model part.

### 6.5.1 Traceability

One important consideration in developing the analysis model is traceability. The analysis model is made up of several diagrams. Sometimes different parts of the model are assigned to different people or done at different times. What is important

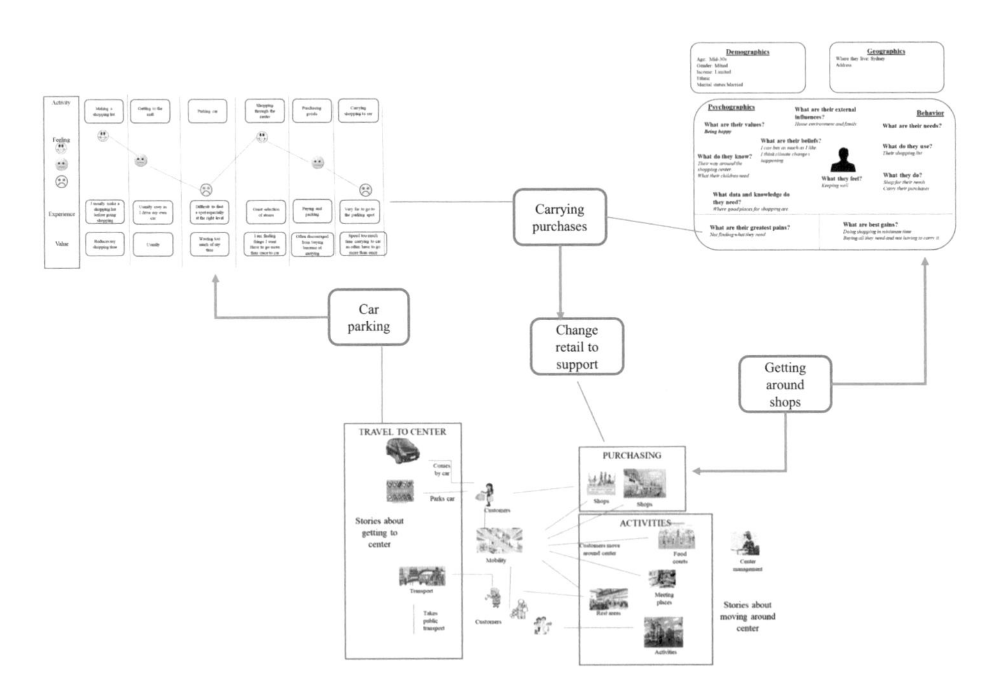


◘ **Fig. 6.6** Showing the complete model

to remember is that all these diagrams and models describe the same system. In that way, it must be possible to look at one part and easily trace its relationship or role in another part—the ability to trace is often called *traceability*. What is often done to ensure traceability, or sometimes called alignment, is to *make sure the same name is used for the same entity everywhere it appears*. Later, in design, solutions will be traced back to the analysis model to ensure that any solution addresses problems found in the analysis.

Traceability can also be used to check the analysis model for consistency. For example, a journey map can be traced to the persona map for the stakeholder.

## 6.6 Identifying Local Issues

Local issues are often indicated by feelings expressed by stakeholders. We can use journey maps to identify stakeholder greatest pains and gains. One goal of any transformation is to remove these pains. The journey map in ◘ Fig. 6.3, for example, shows an unhappy state of a customer carrying purchases to the car, which matches their greatest pain. Finding a car parking space is also an issue, as it spoils an otherwise pleasant trip to the centre.

Once you have a model, you begin to ask questions and identify issues to find out what values need to be improved. Links in the model are important as they define potential value mismatches. For example:

- Do people assigned to a business unit have the necessary skills?

  If not, then finding people with the right skills can become an issue to resolve.
- Are all values of a stakeholder satisfied in their journey maps?

  There are complaints that a process is taking too long.

Designers can now begin to document the mismatches you find as issues. One way is to create an issue note as shown in ◘ Table 6.2 and reasons why stakeholders see it as an issue.

Later, ► Chap. 7 describes how the local issues can be combined with wider global issues to set priorities.

**◘ Table 6.2** A local issue note about local issues

| Question | What are obvious local issues? |
|---|---|
| Stakeholder | What does the stakeholder see as local issue? |
| Customer | – Difficult to carry shopping to the car because of the large distance to the car park |
| Shop owner | – People looking and not buying |
| Centre manager | + Customer base is growing |

## Summary

This chapter described ways to develop models of the system. It showed how the stakeholders work in the system, the relationships between them, and the data they use. It also used journey maps to describe the processes they follow.

## Exercises

For each case study, provide a report that includes the following:

- A summary of the most relevant stories you collected.
- The stakeholders.
- Draw a rich picture.
- The persona maps for three of the most relevant stakeholders.
- Develop journey maps and touchpoint descriptions for these stakeholders.

You can use the following structure in your report.

### Stories

| Story/article | Refers to stakeholder | What is interesting about this story | Keywords |
| --- | --- | --- | --- |
| | | | |
| | | | |
| | | | |
| | | | |
| | | | |

### Stakeholder List

| Stakeholder | What they do | What they value | Way to satisfy value |
| --- | --- | --- | --- |
| | | | |
| | | | |
| | | | |

### Your Rich Picture

Focus on stakeholders, information needs, and relationships.

### Persona Maps for Stakeholders

Any report will include a number of stakeholder persona maps.

**Journey Map of Selected Stakeholders**

Make sure each journey map has a unique name

**Journey Map Touchpoints**

Journey map: Name of journey map

| Touch point name: | Stakeholder-1 | Stakeholder-2 | How is value created |
|---|---|---|---|
| Awareness (finding what is possible) | | | |
| Options (making a choice) | | | |
| Research (what is the best option) | | | |
| Selection and action | | | |
| Services used by stakeholder | | | |

## Further Reading

Curedale, R. (2016). *Journey maps: The tool for design innovation* (1st ed.). Design Community College Inc.

Nooyi, I. (2015, November). How Indra Nooyi turned design thinking into strategy. *Spotlight Interview, Harvard Business Review*.

Rosenbaum, M. S., Otalora, M. L., & Ramirez, G. C. (2017). How to create realistic customer journey map. *Business Horizons, 60*, 143–150.

Tholath, D. I., & Casimirraj, S. J. (2016). Customer journey maps for demographic online customer profiles. *International Journal of Virtual Communities and Social Networking, 8*(1), 1–18.

# Identifying Issues

## Contents




I. T. Hawryszkiewycz, *Transforming Organizations in Disruptive Environments*,
https://doi.org/10.1007/978-981-16-1453-8_7

Once a model of an organization is created, it can be used to identify issues where and why values are not achieved and what needs to create such values in a systematic way. The steps followed are to identify major issues, find their causes, and then to find ways to address these causes.

This chapter goes beyond seeing transformation in complex environments as simply solving a "problem" but addresses a wider challenge—such as addressing issues to improve the health of people in a city, respond to a disruption, or enter a new market in a business.

**Learning Objectives**

- What are issues and problems?
- Negotiation and resolving conflicts
- Ranking issues
- Best practices
- Raising issues based on best practices
- Making decisions



## 7.1 Introduction

The objective of any transformations is to deliver value to people and organizations in changing environments. The need for a transformation often results from changes in people's values, or from natural events or from disruptive innovation. These introduce new challenges, which result in new issues and new problems.

It is important in any transformation to find real problems (Weddell-Wedellsberg, 2017) that address wider issues rather than solve individual or minor local needs that add value to a small number of stakeholders. What is often suggested is to start with an issue seen as important to most stakeholders, and then identify problems that address the issue. Wharton School of Management or organizations like Mediate, a consulting company, also see identifying real problems as important. By addressing wider issues, transformations can deliver value to many stakeholders. The general trend is to identify issues that affect many stakeholders, then identify their causes, and then identify the problems as removing their causes.

It is almost inevitable that transformation will lead to changes to the way a business works. Such changes will affect many stakeholders. In many cases, change can result in conflicts, especially where some stakeholders' values are affected negatively, while others are affected positively. Such conflicts (Liddle, 2017) need to be negotiated and resolved until all stakeholders are satisfied (Jabri, 2017; Johnson, 2017) often in global situations (Maude, 2014). Often outcomes must be such that no stakeholders see themselves as losers. Thus, the tools proposed here not only focus on solving the technical and organizational structures but also on personal value issues.

- Agreeing on issues seen as important to stakeholders.
- Finding ways to address the issues and set priorities.
- Identify potential conflicts in stakeholder values.
- Negotiating to resolve the conflicts.
- Making decisions acceptable to all.

Stakeholder values play an important role in any negotiation and decision making. The chapter will describe how to rank these problems in priority order acceptable to the stakeholders. The chapter begins by describing what issues are, then how to find their causes, and then focuses on finding problems by identifying where best practices are not met.

## 7.2 What Are Issues?

Issues are defined in many ways. In the broadest sense, they can relate to what everyone is talking about. Is it loss of a major client? Or is it introduction of a new product? Or, in the broader sense, how to address climate change? Or is it what to do in a pandemic? We often gather stories about what is going on by talking to people and then raise issues from what we find in the stories. Another definition is that an issue is something that everyone is talking about.

Issues can be local or global. Local issues are internal to your business—for example, there are often delays at some point of our service delivery system. They are often caused by some local problem and often locally resolved. Global issues, on the other hand, are those that result in changes or expectations in our environment. They can be disruptions in the global supply chain of which we are a part. They can be pandemics, like COVID-19. Increasingly now, because of globalization, global issues can have sudden local impact and their causes must also be locally addressed, as for example, by finding an alternate way to that supply chain.

Then there are issues that arise from external factors. Often, they relate to business values—for example, we are losing customers, our services are uncompetitive. They can be more general such as:

- An important topic for discussion raised by an influential stakeholder.
- Resolving a matter of dispute between parties.
- New government regulation.

They can also be a general statement such as "poor reservoir and land management have revealed a lack of long-term planning on these *issues*," the experts have said.

Issues also arise in cities, as for example, "safety needs to be addressed especially in a selected area." It is often difficult in informal discussions to prioritize issues to allocate resources in ways that optimize outcomes.

## 7.3 How to Find Issues

The first step in raising an issue is to define the context—a particular business or city, for example. Context can be narrow or wide. The context can be part of a business or city. They can also be as simple as planning a holiday or difficult, such as how to reduce the level of pollution in a city. Ultimately, designers must raise those issues that have most impact on value for both people and organizations. A common way in many organizations is to use raise issues by comparing what is happening now to best practices. These ask questions about what is important to an organization, or what is important to a business.

### 7.3.1 A Systematic Approach to Finding Issues

There are often many issues to be addressed in a transformation. If you look back at today's problems as a set of tangled strings, and our goal is to untangle them. Each knot is an issue, but you cannot simply undo one knot by removing one string from a tangle—you need to take a little bit of one, and then another, and so on. Often, if you grab one quickly and pull it tight, you might find yourself creating new knots. You also learn as you go along—often seeing new knots and unwinding them. Similarly, issues in business or cities must be addressed in a systematic manner—a bit at a time, or a problem at a time. Thus, adding a sensor to open doors at the same time can contribute to city safety.

Figure 7.1 illustrates activities to identify problems in a systematic manner. The process involves stakeholders in all activities, whereas the management organizes the activities and assigns stakeholders to them.



> Any activities should use methods and tools to identify problems important to all, initiate in-depth discussion, resulting in productive discussion and negotiation and lead to decisions acceptable to all.

Figure 7.1 shows the activities followed to achieve a successful outcome.

The activities call for continuous stakeholder participation, where stakeholders contribute to the discussion and negotiation from their value perspective. The activities in Fig. 7.1 are as follows:

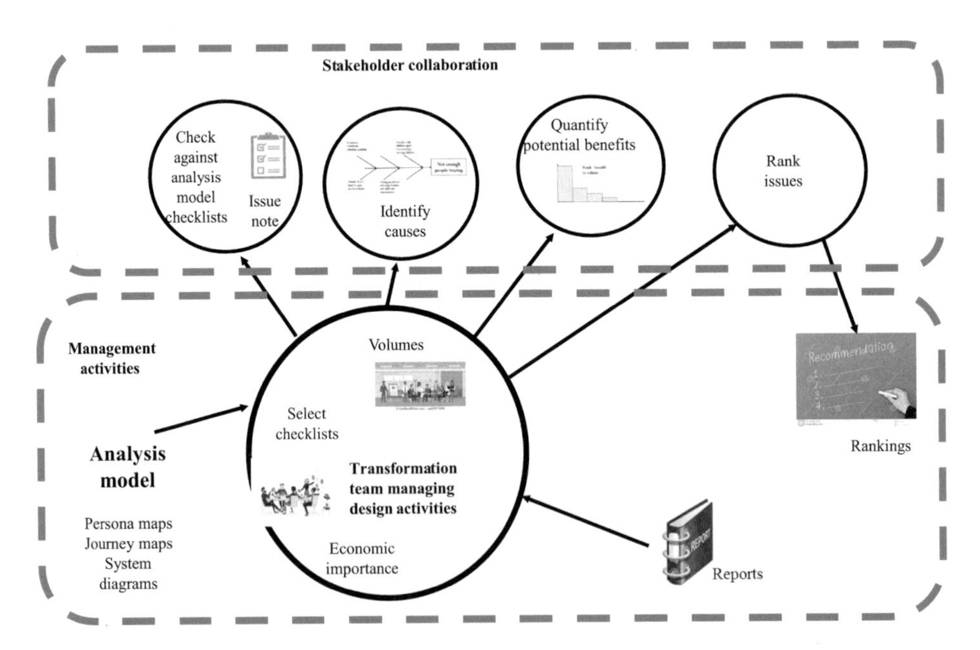


Fig. 7.1 Activities in issue analysis

- Select and/or develop a checklist based on industry standards, raise checklist questions, and identify potential issues and stakeholder attitudes to them, and raise an issue note.
- Look for causes of an issue.
- Seek stakeholders input on the importance of an issue, and benefits to the stakeholder.
- Rank issues based on stakeholder values to set priorities and see which should be addressed first.

These are coordinated by a leadership group that brings together teams with the skills needed in each activity.

## 7.4 Selecting Checklists

Checklists are commonly used in practice as a first step. Checklists are based on best practices and challenges faced in industry. These best practices were introduced in the previous chapter. Checklists in most cases are a set of questions on whether values, such as those defined in ▶ Chaps. 3 and 4, are being met. A checklist is simply a set of questions. For example, are you doing the best possible to retain clients and to grow your client list? We ask each question in the context and note stakeholder's response.

One early step is to agree on the checklists to be used. Once a checklist is chosen, it is used to evaluate impact on values, by asking stakeholders about their response to checklist questions. Where stakeholder views cannot be obtained, the design team can use the previously developed analysis model, which includes stakeholders' values and feelings, to evaluate their possible response. Ultimately, the outcome is a list issues seen as important.

### 7.4.1 Business Value Checklist

Checklist for business value is shown in ◻ Table 7.1. It is derived from ▶ Table 4.2, which defined business values. Each of the characteristics of a business value has been converted to a set of questions. These questions are posed to see if the business value is satisfied. The questions focus on these business values. These, as earlier defined in ▶ Chap. 4, are divided into four themes, namely, customer satisfaction, developing a revenue base, staying relevant, and with strong internal performance.

◻ Table 7.1 focuses on what are historically important values in business, especially in survival in the commercial sense. There are now other values emerging. These arise from sustainability issues often arising from climate change and increasing disruption.

The best practices in ◻ Table 7.2 see resilience and agility as important. Resilience is becoming an increasingly important topic in climate change and resilience to disasters. The incidence of disasters is now growing, and the range of disasters will be discussed in greater detail in ▶ Chaps. 9 and 10. It is, however, important to remem-

**Table 7.1** Checklist for business value

| Business value themes | Questions to be asked to find unsatisfied values |
|---|---|
| Customer satisfaction<br>Which of these add to your satisfaction about customer satisfaction? | Are customers coming back?<br>Is your customer base growing?<br>Is there a loyalty program?<br>Is access to your business easy?<br>Is your business close to public transport?<br>Do you have a website?<br>Is there a well understood public mission?<br>Are you getting new customers?<br>Do you stay in touch with customers?<br>Do you provide feedback to customers?<br>Can you easily explain how your services and products have developed? Are they safe to use? |
| Developing revenue base for financial success<br>Does this activity add to revenue? | Is revenue growing?<br>Do you have good decision-making processes?<br>Are your costs reducing?<br>Is outsourcing growing and providing benefits?<br>Are you getting value from technology?<br>Do you have a financial plan?<br>How do you make investment decisions?<br>Are predictive analytics and social media used?<br>Do you respond quickly to unexpected incidents? |
| Staying relevant<br>Are you satisfied that business activities are relevant to stakeholders? | Are products and services continually improving?<br>How do you evaluate the benefits of technology?<br>Do you listen to what people, including customers, are saying?<br>Do you keep track of similar products?<br>Do you collect and analyse feedback?<br>Do you have an innovation program?<br>How do you build your reputation?<br>Are you finding new markets?<br>Do you have a process to respond to a new competitor? |
| Growth | Are you changing your products in line with customer needs?<br>Does your new development satisfy green criteria?? |
| Internal performance | Are your employees happy?<br>Is there a large turnover? Is there a large turnover of employees?<br>How do you develop employee skills?<br>Do your employees contribute to new directions?<br>Are your employees agile?<br>Do you quickly correct problems in your organization? |



ber here that disaster can have a significant impact on businesses and cities. These must have the infrastructure to respond to disasters and their checklists include questions of how to respond to disasters. Major disasters such as pandemics or floods often require greater global response and ways to assemble communities to deal with them. Ways to deal with disasters is described in more detail in ▸ Chap. 9.

**Table 7.2** Emerging issues

| Environmental value themes | Questions to be asked to identify unsatisfied values |
| --- | --- |
| Business resilience | Do you have a risk analysis program?<br>Do you continually scan the environment for threats?<br>Do you resource redundancy program?<br>Is it easy to change management responsibility? |
| Agility | Do you have ways to respond to change?<br>Are your teams empowered to respond to change?<br>Can resources be quickly reassigned?<br>Do you have a stress testing activity? |
| Waste management | Do you have a policy for waste disposal? |
| Energy conservation | Is there a policy for energy conservation? |
| Emission standards | Is there a policy for minimizing emissions?<br>Is your use of reusable energy growing?<br>Do you measure your carbon footprint? |

### 7.4.2 City Value Checklist

The same applies to a checklist based on city values, which is shown in Table 7.3.

### 7.4.3 Customized Checklists

The checklists in Tables 7.1, 7.2, and 7.3 are common to most businesses and cities. Checklists can be developed for any context. Many people, for example, have a checklist when they travel. They know what they need depending on where they go. When you find that you do not have something it becomes a problem, although minor. The cause is that you forgot to buy it, and the solution is to go and buy whatever is missing and thus solve the problem.

## 7.5 Raising Issues

The book uses the guidelines introduced earlier for identifying real problems, especially involving all concerned stakeholders, and focusing on best practices. The process for raising issue notes from checklists is shown in Figs. 7.2 and 7.3. The book provided some checklists earlier in this chapter. Table 7.1, for example, showed a checklist for businesses, Table 7.3 is a checklist for cities.

One oft-used way is to use questions in the checklist of best practices. Then, for each question, get a value-based judgement from affected stakeholders; a plus for agreement or a minus otherwise. The value-based judgement can of course be supported with data. A question that has negative comments is then flagged or raised as an issue.

**Table 7.3** City checklist

| City value themes | Questions to ask to find unsatisfied values |
|---|---|
| Walkability<br>What is your answer to? | Are there wide streets and open spaces?<br>Does city support movement of pedestrians rather than cars?<br>Are there spaces to rest or take breaks or meet people—for coffee?<br>Are there things to see while walking?<br>Are streets safe for pedestrians?<br>Is there easy access to public transport?<br>Is there a centre?<br>Are people using public transport?<br>Are there plenty of spaces to play on?<br>Are streets designed for bicycles and pedestrians?<br>Are places of interest close together? |
| Safety<br>What is your answer to? | Has there been a reduction in crime levels?<br>Is there adequate sanitation?<br>Are there safety instructions in the event of a disaster?<br>Is the level of pollution acceptable?<br>Is there a safe city policy (see Singapore, London, Dubai)?<br>Can you walk alone at night?<br>Are services available if injured or sick?<br>Are all segments of citizens safe?<br>Are there monitoring devices?<br>Are predictive analytics and social media used?<br>Is response to incidents quick? |
| Health<br>What is your answer to? | Is there adequate sanitation?<br>Are there less cars and more trees?<br>Is there access to clean water?<br>Is there access to good food?<br>Is slum housing going down?<br>Is air quality improving?<br>Is urban violence dropping? |
| Environment<br>What is your answer to? | Is there a feeling that the city is part of nature?<br>Are there shared outdoor spaces that promote physical activity?<br>Are there more trees today than earlier? |
| Mobility<br>What is your opinion on? | Can you quickly get to shops to buy necessary goods?<br>Is it easy to get your purchases to your home?<br>Can you quickly get from one place to another? |
| Connectivity | Is it easy to find access to services? |
| Housing | Is there sufficient affordable housing? |
| Vibrant community spaces<br>What is your answer to? | Are there meeting places to share knowledge and ideas?<br>Are there safe spaces where people can gather at any time? |
| Sustainability<br>What is your answer to? | Sharing resources (public transportation).<br>Are alternate energy sources increasing?<br>Do apartments have no lawns—reduced watering. |
| Use of data and data analytics.<br>What is your answer to? | Is city-wide data collected?<br>Is the data analysed to make the city smarter?<br>Is the data specific to an application or generally open and available? |

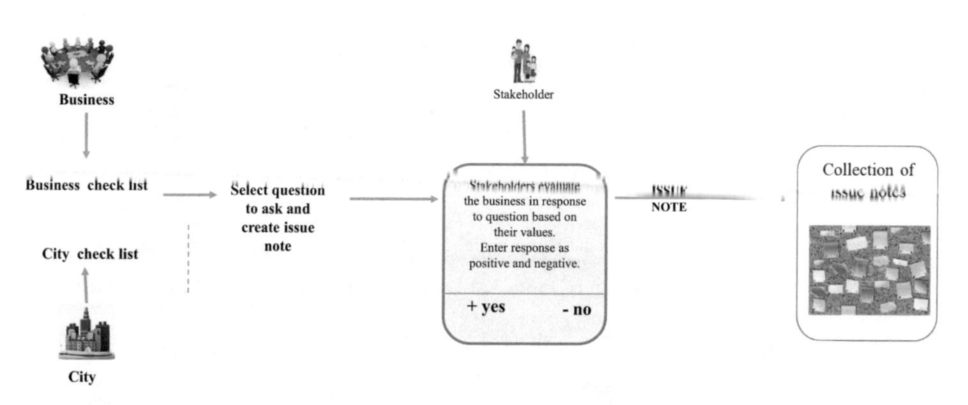


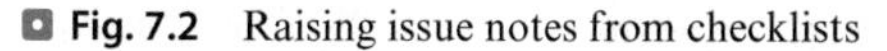

**Fig. 7.2** Raising issue notes from checklists

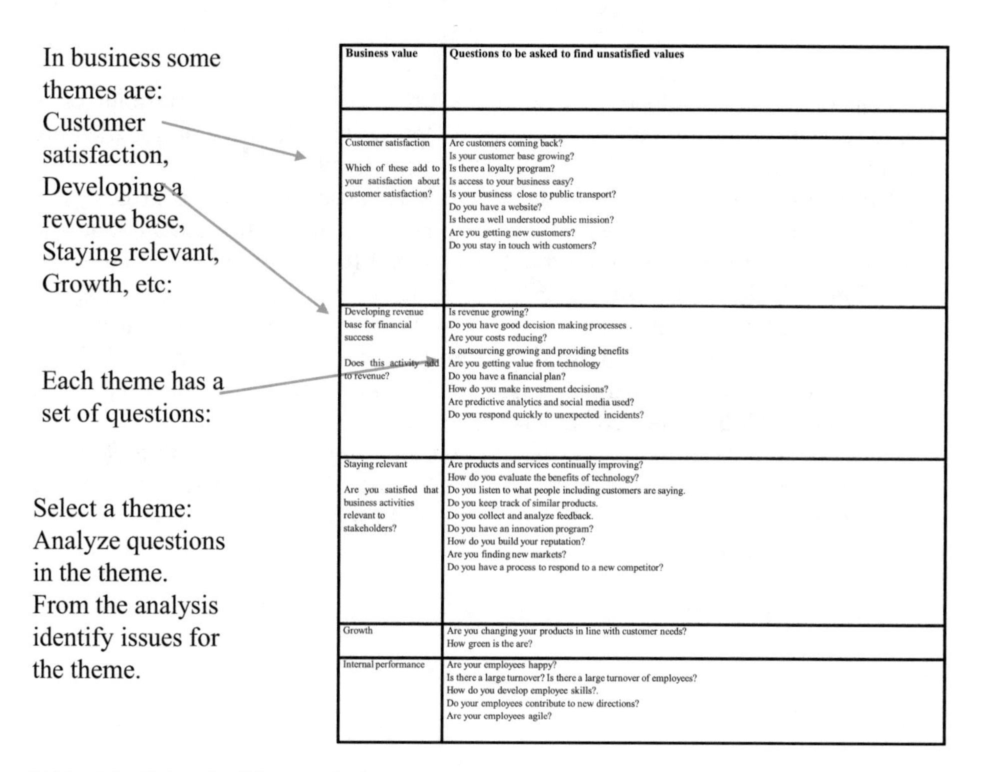


| Business value | Questions to be asked to find unsatisfied values |
|---|---|
| | |
| Customer satisfaction<br><br>Which of these add to your satisfaction about customer satisfaction? | Are customers coming back?<br>Is your customer base growing?<br>Is there a loyalty program?<br>Is access to your business easy?<br>Is your business close to public transport?<br>Do you have a website?<br>Is there a well understood public mission?<br>Are you getting new customers?<br>Do you stay in touch with customers? |
| Developing revenue base for financial success<br><br>Does this activity add to revenue? | Is revenue growing?<br>Do you have good decision making processes .<br>Are your costs reducing?<br>Is outsourcing growing and providing benefits<br>Are you getting value from technology<br>Do you have a financial plan?<br>How do you make investment decisions?<br>Are predictive analytics and social media used?<br>Do you respond quickly to unexpected incidents? |
| Staying relevant<br><br>Are you satisfied that business activities relevant to stakeholders? | Are products and services continually improving?<br>How do you evaluate the benefits of technology?<br>Do you listen to what people including customers are saying.<br>Do you keep track of similar products.<br>Do you collect and analyze feedback.<br>Do you have an innovation program?<br>How do you build your reputation?<br>Are you finding new markets?<br>Do you have a process to respond to a new competitor? |
| Growth | Are you changing your products in line with customer needs?<br>How green is the are? |
| Internal performance | Are your employees happy?<br>Is there a large turnover? Is there a large turnover of employees?<br>How do you develop employee skills?.<br>Do your employees contribute to new directions?<br>Are your employees agile? |

**Fig. 7.3** Using checklists to raise issues

*Select one of the questions in the checklist*, for example "Are customers buying?"

Then *get a value judgement from different affected stakeholders*.

Now, issue notes like that in Table 7.4 can be created to record stakeholder value on a scale of 1 to 7. These value judgements record each stakeholder's view on whether their needs are or are not met.

**Table 7.4** Issue note on "Are customers buying?"

| Checklist: | Business value | |
|---|---|---|
| | Theme: Customer satisfaction<br>Question: Are customers buying? | |
| Stakeholder | Stakeholder value judgement on question—do stakeholders agree (7) or not (1)? | |
| Customer | 3 | No, because it is getting hard to get shopping home if you come by public transport. |
| | 6 | Good selection of things I need |
| Shop owner | 4 | Most people are looking and not buying |
| Centre Manager | 5 | There should be more sales. Our figures show a 5% growth of people entering the shopping centre. |



For a business, questions focus on customer satisfaction, developing a revenue base, staying relevant, and internal performance. What designers often look for then are common features or themes. Often sorting helps to identify them.

Table 7.4 is an example of an issue note. It is not a standard issue note (there is no such standard), but shows the kind of information needed to identify issues to address.

It shows stakeholder evaluation to a question as 1 (disagree) to 7 (fully agree).

Obviously dealing with a huge number of issue notes makes it difficult to suggest any solutions or recommendations. Ultimately these must be organized to lead to decisions. Brainstorming now must also be organized. It is not getting people in an office together and have a general discussion. The brainstorming, on the other hand, needs to look at all the collected issue notes.

Sometimes an in-depth analysis is required to present concrete data to support a value-based judgement before any decision to raise the issue made. Initial issue analysis can often be distributed to individual team members—at least to small groups. In large systems, we would plan for each team member to take one business value.

An in-depth analysis can be based on the analysis model.

### An In-Depth Analysis Based on the Analysis Model

Look at Journey maps. It is possible to compare each touchpoint in the journey map to persona maps to identify where value is not met.

Brainstorming can ask questions like what is the cause of less people buying in our shopping centre—is it because of our infrastructure (difficulty of going from store to store) or our limited range of shops?

Going back to the challenges described in ► Chap. 4 also gives a guideline for questions. The elderly not buying because of difficulty of carrying shopping home. Is it families with lack of time caused by minding children?

Perhaps given the shopping mall goal is to focus on survival of the mall. A common theme is not of people buying at one shop but the fact that more people are visiting the mall but not visiting shops.

Why is there reduced customer satisfaction? Why are they not buying? What are the causes? The question then becomes "is there sufficient evidence to suggest that customer satisfaction is a major issue?" All that is possible is to generate summaries and relationships and make decisions through brainstorming.

In summary, issue notes should be an organized way to discuss and systematically rank issues and problems that follow. The next step in ◘ Fig. 7.1 is to analyse issues and identify issue causes.

## 7.6 Issue Analysis—Causes and Importance

A *fishbone diagram* like that shown in ◘ Fig. 7.4 sometimes helps to find causes. The issue is on the right-hand side and through brainstorming designers identify what the potential causes are. In the case of ◘ Fig. 7.4, one cause may be that families with children spend too much time minding children rather than shopping.

*Pareto diagrams* can then be used to show the importance of each cause. Ranking the causes—here we might look at which of these impact on what customer segments. We can then analyse the importance of each cause.

These are often seen as what if questions? What if we develop a website? How many people will benefit? Some quantitative data is often needed to compare alternatives (◘ Fig. 7.5).

### 7.6.1 Recording as Issue Notes

It is always a good idea to keep track of issue notes and trace back to the stakeholders and journey maps that cause the issue. Issue notes provide a way to keep such records by recording of how issues impact on stakeholder values (◘ Fig. 7.6) and the change after a proposed transformation.

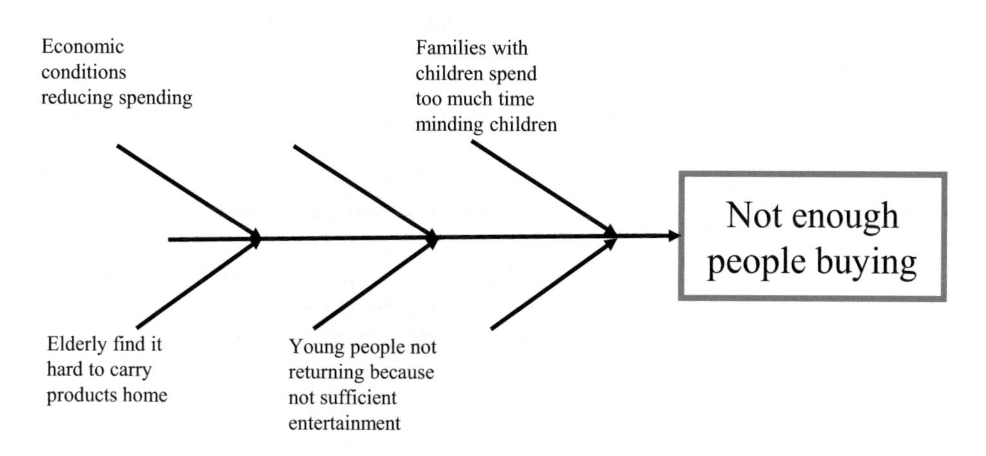


◘ **Fig. 7.4** Fishbone diagram showing causes of issues

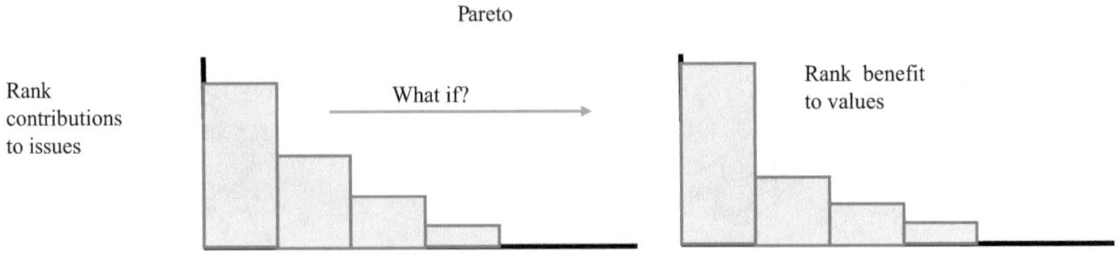
Pareto
Rank contributions to issues
What if?
Rank benefit to values

**Fig. 7.5** Evaluating importance

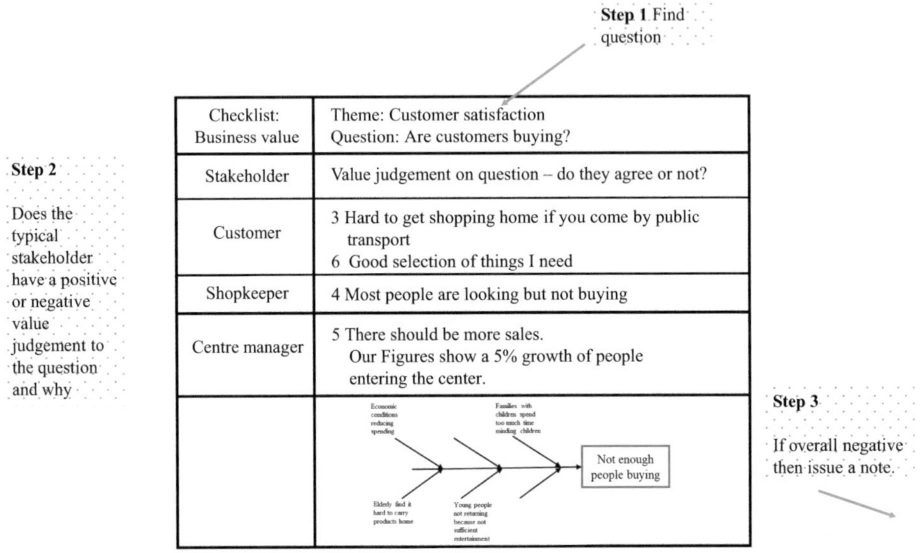
Step 1 Find question
Step 2
Does the typical stakeholder have a positive or negative value judgement to the question and why
Checklist: Business value
Theme: Customer satisfaction
Question: Are customers buying?
Stakeholder
Value judgement on question – do they agree or not?
Customer
3 Hard to get shopping home if you come by public transport
6 Good selection of things I need
Shopkeeper
4 Most people are looking but not buying
Centre manager
5 There should be more sales.
Our Figures show a 5% growth of people entering the center.
Economic conditions reducing spending
Families with children spend too much time minding children
Not enough people buying
Elderly find it hard to carry products home
Young people not returning because not sufficient entertainment
Step 3
If overall negative then issue a note.

**Fig. 7.6** Issue note now with fishbone diagram added

### 7.6.2 Evaluating Groups of Issues

There are often many questions in a checklist, which in turn can lead to many issue notes. One way to minimize the number of issue notes is to look at groups of questions as a theme. For example, customer satisfaction as a theme, has a number of best practice questions. In that case, we construct a table like Table 7.5 that has a separate column for each question in customer satisfaction theme and how different stakeholders respond to it.

We can also compare the different themes. Table 7.6 shows the evaluation of each business value theme by each stakeholder. Business customer segments to create value to both.

Thus, initially we select recommendations based primarily on value considerations rather than cost issues. These are considered next in ▶ Chap. 7, where we look at ways to integrate the recommendations into the existing system.

Table 7.5 Are customers satisfied with the shopping centre?

| Checklist | Business value | Business value |
|---|---|---|
| Question | Theme: Customer satisfaction:<br>Question: Is access to your business easy? | Theme: Customer satisfaction:<br>Question: Do you stay in touch with customers? |
| Stakeholder | Value judgement on question—do stakeholders agree or not? | Value judgement on question—do stakeholders agree or not? |
| Customer | 4 Tends to be crowded | 2 No, I do not remember the shopkeeper following up. |
| Shop owner | 4 Need more space for customers to browse | 4 I try to keep products people like by constantly getting feedback. |
| Centre Manager | 6 Our cost controls are paying off. Introduce new services to make shopping easier. | 5 We advertise in the local press. |

Table 7.6 Is the shopping centre meeting business values?

| Checklist | Business value | Business value |
|---|---|---|
| Question | Theme: Customer satisfaction: | Theme: Maintaining strong internal performance |
| It is then possible to find common causes stakeholder | Value judgement on question—do stakeholders agree or not? | Value judgement on question—do stakeholders agree or not? |
| Customer | | |
| Shop owner | 6 Revenue base increase require more space | 5 I try to keep products people like by constantly getting feedback. |
| Centre Manager | 5 Our cost controls are paying off. Introduce new services to make shopping easier. | 6 We are lacking in entertainment.<br>7 Lack of support for elderly. |

### 7.6.3 Finding the Causes of Issues

Grouping by theme can also result in a new fishbone diagram for each question. It then makes it possible to identify common causes for each question in a selected theme. Figure 7.7, for example, shows some causes of issue notes in the customer satisfaction theme.

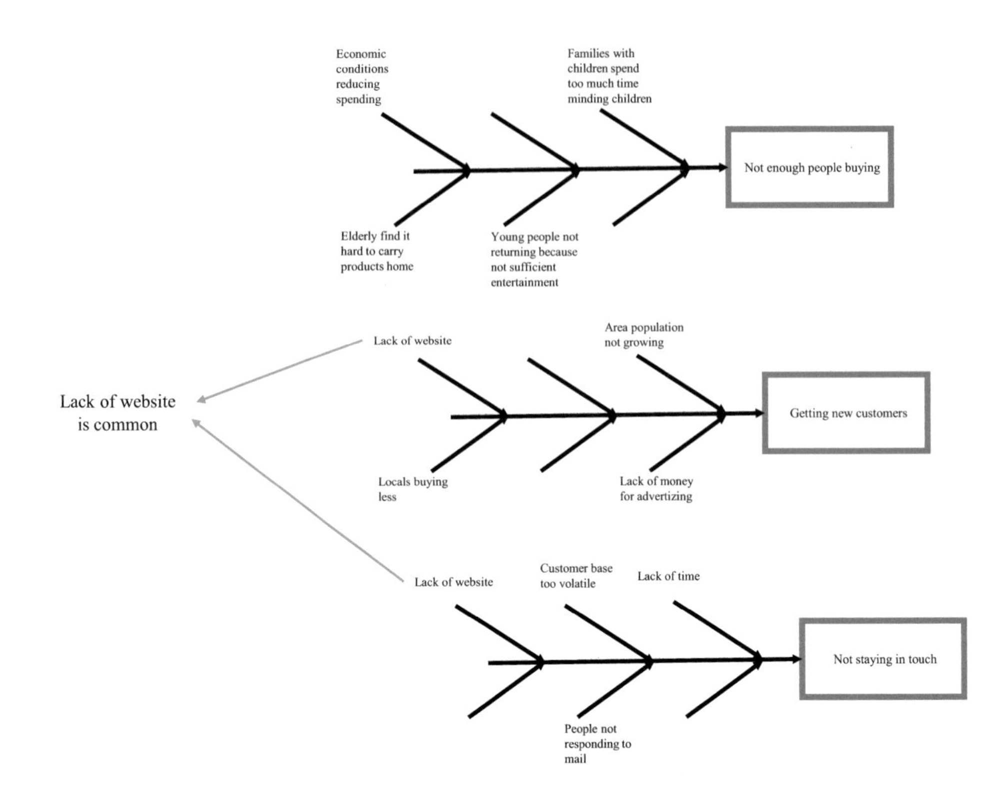


**Fig. 7.7** Looking for holistic solutions

**Table 7.7** Selected issues

| Worthwhile issues | Customer focus | What needs to be done |
|---|---|---|
| Issue 1: Public transportation inconvenient | Elderly who need public transport | Possibly add to the car park |
| Issue 2: Difficult to get shopping home | Most customers who make large purchases | Arrange for deliveries to be made |
| Issue 3: Insufficient centre facilities | Young who want more entertainment. Also, families, but who also need child minding. | Provide entertainment and child-minding facilities. |

Brainstorming is the most oft-used method to generate ideas. Brainstorming is often seen as coming with ideas in an abductive way—asking what if questions. Or by asking what the cause of a problem is and what if we find ways to remove this cause.

In a business, there are many issues; it becomes impossible to address them all. Consequently, some issues are selected or prioritized often on political grounds. Designers might then create a table that shows selected issues for a theme. Table 7.7 is an example of the contents of a table that describe the issues and what can be done to address them.

## 7.7 Making Decisions

In any environment, there are always many issues, some are important, whereas others are less important. Ultimately, a decision must be made on what issues to focus on. Again, we come back to focus on those that are real problems by stakeholders. Thus, issue notes indicate what values stakeholders place on issues. At the same time, it is necessary to propose problems that are possible to solve and provide value benefits.

One term often used to combine these is the "joint value proposition".

### 7.7.1 Joint Value Proposition

The joint value proposition basically says <we will do X> so that < value p will improve> for <stakeholder-a> and <value-q> will improve for <stakeholder-b>. For example:

We will increase the variety of products in the shop so that customers benefit because they find more things to buy, and shopkeepers benefit by selling more products. It is important in formulating the joint value proposition that:

- Any value gains address the values in the persona map, especially the pains and gains.
- As many people benefit as possible.
- The value judgement in the issue notes is addressed.

In addition, any proposal should be reasonably achievable. Eventually it calls for a detailed evaluation, including cost benefit analysis, identify any process changes, data, and technology needs. It is possible now to rank potential solutions and select which of these to evaluate in detail.

### 7.7.2 Ranking Issues

The next step is to rank issues to see which to focus on. Ranking of issues, ideas, and what to do next inevitably leads to lengthy discussion arising from conflicts (Liddle, 2017). Often conflict can be avoided by first adding wider organizational criteria (rather than just using stakeholder values) for ranking. The process follows the steps as follows.

Step 1: Identify organizational criteria to be used in the evaluation, what is seen as important to the business. These can be set by the design team in ◘ Fig. 7.1.
Step 2: Collect the evaluation of issues based on the criteria from stakeholders.
Step 3: Rank issues against business criteria.

The idea is that designers, rather than focusing on comparing issues themselves, first define the criteria on which they are ranked. These usually focus on what are core

Table 7.8 Ranking issues

| Issue Growing customers | A—Current stakeholder value (1—low, 7—high) | B—Will most stakeholders benefit (7 yes, 1 no) | C—The difficulty of making the proposed change (1—easy, 7—hard) | D—Expected cost (1 low, 7 high) | Summary (A + B)/(C + D) |
|---|---|---|---|---|---|
| Option What can be done | | | | | |
| Setting up child minding | 4.7 | 3 | 3 | 4 | 1.1 (4.7 + 3)/(3 + 4) |
| Providing entertainment | 2 | 2 | 4 | 7 | 0.36 |
| Increase shopping hours | 3 | 6 | 3 | 4 | 1.3 |

values in organization. Negotiation can focus more on the ranking of an issue on the business values rather than stakeholder values. Thus, for example, in Table 7.8.

Chosen Evaluation Criteria in Table 7.8

A. Stakeholder values what is important to stakeholders.
B. The likelihood of improving values of all stakeholders.
C. Difficulty of making changes because of extensive needed organizational structures.
D. Cost difficulty.

In Table 7.8, criteria A and B, the higher the better; in the cases of criteria C and D, the higher. The criteria will then probably require some evaluation, such as making cost estimates.

Then each criterion is ranked starting with 1 as the easiest. We then add all the rankings and find the easiest change; in Table 7.8, the easiest way is to extend the hours of shopping.

In summary, we divide the benefits (Ranking in A and B) by the difficulties (Ranking in C and D)—the higher the number, the more likely that the option be adopted; the higher the difficulty, the less likely for the option to be adopted.

Here, increasing shopping hours is seen as the easiest way to increase customers as it will benefit many stakeholders and is relatively easy to implement as it does not require any physical rearrangement of facilities.

Should some people disagree, the argument will be not that I like another option better, but to argue about the criteria—for example, we could lower the cost of child minding by businesses making free space available.

## Summary

This chapter and the previous chapter described the issues and challenges faced by businesses. It described various methods used to identify issues. Although these were described in sequence, in practice, designers or organizations may not follow such sequence, use similar methods, or have their own customized methods. Nevertheless, all such methods have the same goal—to identify issues to address and problems to solve to get the best value.

This chapter also described ways in which conflicts that frequently arise when making decisions. It was to use tools that focused more on broader criteria rather than those on specific issues such as checklists, fishbone diagrams, and then on organization-wide criteria when selecting issues to address.

## Exercise

In doing so, encourage your teams to follow the steps in ◘ Fig. 7.1 as a set of brainstorming sessions. For your best practice questions select the following:

Q1: Can you easily explain how your services and products have developed? Are they safe to use?

Q2: Do you quickly correct problems in your organization?

Q1, for example, applies to both restaurants and supply chains. Is it safe to eat a meal? Do you know whether its food has been sourced from a safe location? Has it been prepared safely? For a supply chain, is it possible to trace where a product has come from, in say a supermarket?

Q2 also applies to both. In a supply chain, there may be a problem in deliveries in some part of the supply chain. In a restaurant, there may be a fault in an oven or a late delivery of products.

You can use the following template to record the outcome of your brainstorming.

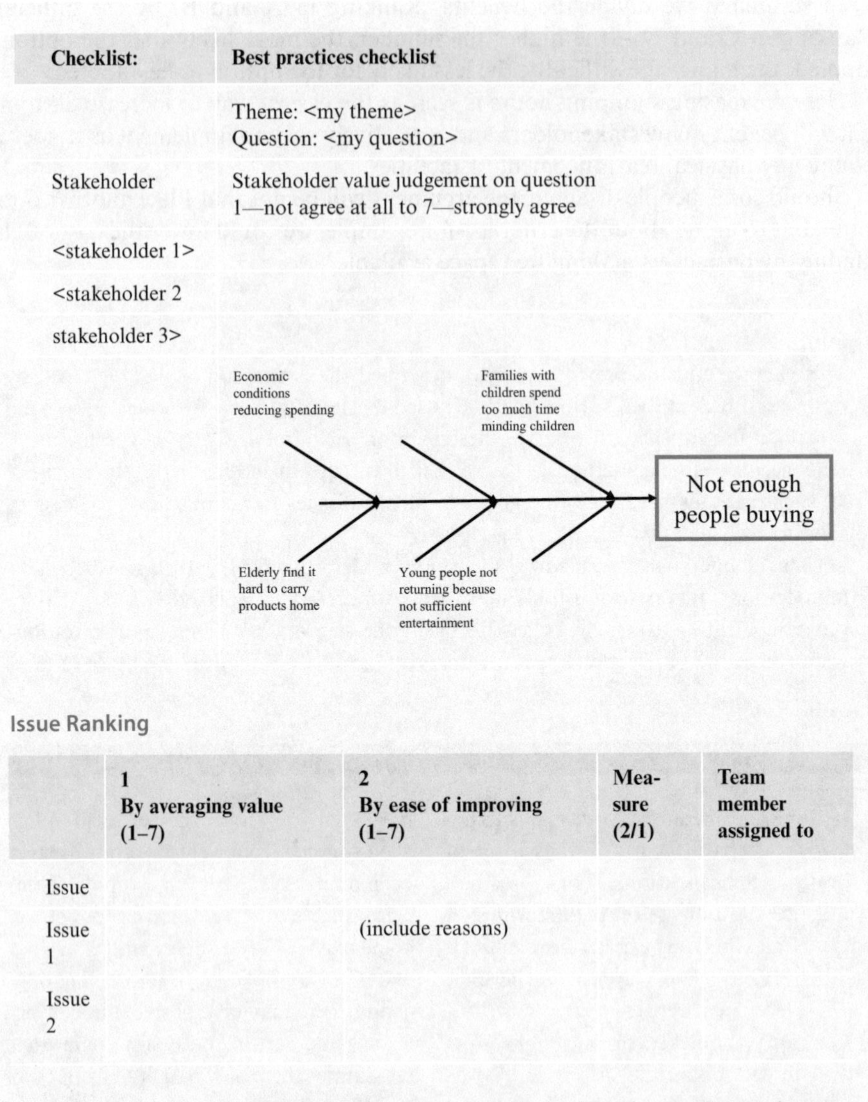

| Checklist: | Best practices checklist |
|---|---|
| | Theme: <my theme><br>Question: <my question> |
| Stakeholder | Stakeholder value judgement on question<br>1—not agree at all to 7—strongly agree |
| <stakeholder 1> | |
| <stakeholder 2 | |
| stakeholder 3> | |



## Issue Ranking

| | 1<br>By averaging value<br>(1–7) | 2<br>By ease of improving<br>(1–7) | Mea-<br>sure<br>(2/1) | Team<br>member<br>assigned to |
|---|---|---|---|---|
| Issue | | | | |
| Issue<br>1 | | (include reasons) | | |
| Issue<br>2 | | | | |

You may wish to develop customized checklists by looking through Google to identify issues such as transparency, reliability of supply to the consumer, safety, communication, rising costs, and inventory. Would these issues change for people with special needs. Extend best practices to address pandemic, such as keeping food clean in delivery.

## References

Jabri, M. (2017). *Managing organizational change*. Palgrave Macmillan.

Johnson, K. M. (Ed.). (2017). *Negotiation: Insights, strategies and outcomes*. Nova Science Publishers.

Liddle, D. (2017). *Managing conflict*. Kogan Page Ltd.

Maude, B. (2014). *International business negotiation*. Palgrave Macmillan.

Weddell-Wedellsberg, T. (2017, January–February). Are you solving the right problems. *Harvard Business Review*.

## Further Reading

Coleman, S., & Thomas, B. (2017). *Organizational change explained*. Kogan Page Ltd.

Harmon, P. (2019). *Business process change*. Morgan Kaufman.

Liedtka, J., King, A., & Bennett, K. (2013). *Solving problems with design thinking*. Columbia Press.

Mediate. https://www.mediate.com/articles/thicks.cfm (Last accessed: November 27, 2020).

Wharton School of Management. (2015). https://executiveeducation.wharton.upenn.edu/thought-leadership/wharton-at-work/2015/06/identify-the-real-problem/ (Last accessed: November 27, 2020).

# Creating Solutions

## Contents




I. T. Hawryszkiewycz, *Transforming Organizations in Disruptive Environments*,
https://doi.org/10.1007/978-981-16-1453-8_8

So far data about the organization has been collected and a model of the organization developed. Issues important to stakeholders have also been defined. Now we begin to create solutions to address these issues.

Solutions are created by brainstorming and reasoning to generate innovative ideas that address issues and generate ways to implement these ideas. Such ways are described by changes that are made to the model.

**Learning Objectives**

- Reasoning to identify transformations
- Brainstorming
- Joint value proposition
- Identifying data needs
- Identifying technology needs
- How to encourage creativity

## 8.1 Introduction

The previous chapter described ways to identify the issues to be addressed in a transformation. The next step is to provide solutions to address these issues. Here designers look for ideas on how to transform the system and ways to put them into practice.

Solutions can emerge in different ways. This chapter describes two ways that solutions are reached in practice.

- A *solution-based approach* is to adapt a previously used method to provide a solution. Start-ups often start with a method or technology, which is then adapted to a particular business or industry need.
- A *value focused approach* that first identifies needs as described in ▶ Chap. 7, and then delivers solutions to satisfy these needs.

There are of course other ways in which innovation emerges. One is where a known technology is developed to improve a specific task. After all, replacing paper-based bookkeeping by a computer using Excel was an innovation to many businesses—instead of working with pen and paper over books, they now had the time to do other things to improve their business.

Another characteristic of innovation is that it often transcends application. Often, once a new way of using a technology is developed, people discover new applications for it. The iPhone, for example, started from the iPod satisfying people's need for music. Once adopted, new applications were incorporated to the extent that now it is an essential part of many people's lives. Thus, once any innovation is adopted, it is always worthwhile to see how it can be extended into other possibilities to meet people's emerging needs.

Many successful businesses started simple but grew relatively rapidly once value is realized. An example here is Uber. Uber started to connect people in San Francisco in travelling to work. Once the value was realized, its extension beyond putting known people together to the wider community became obvious. All it needed was investment to create the technology platform and develop data on people who want to use the service. It demonstrated that one way to grow is to create a common good, that is, to provide value to all stakeholders in the business.

### 8.1.1 Delivering a Common Good

One common set of criteria in any innovation is that concerned with delivering some common good (Dorst et al., 2016), that is, to create value to all stakeholders.

- Focus on the core of the business or process.
- How to improve activities, not changing them.
- The solution should focus on what people want to be in the future.
- Clearly defining what the core business is.
- Making sure that any design must contribute to the core business.



## 8.2 What Is Important in Design

Successful transformations usually result when stakeholders collaborate by discussing or brainstorming issues (García-García, 2017). They ask questions such as "What if we do something?" Careful and logical reasoning is needed to make a good choice in design.

- What is needed to develop solutions is:
- Collaborative brainstorming to develop ideas.
- Reasoning on how the ideas can satisfy all stakeholders.

Brainstorming, which includes a range of community stakeholders, is to find ideas and resolve issues in ways that add to stakeholder value. In most cases, ideas emerge through brainstorming, followed by discussion where people familiar with the situation can reason in a systematic way on the best practical ideas that lead to solutions. Reasoning is used in brainstorming to identify how ideas can lead to acceptable solutions.

The goal of collaborative brainstorming is to encourage creativity, and critical thinking results in ideas (Taura & Nagai, 2017) where, often intuition and experience suggest solutions that must then be evaluated by the design team. It requires a balance between people with expert knowledge and what is called out-of-the-box or divergent thinking (An et al., 2016) to lead to discussions that result in creative outcomes.

The kind of processes followed in design thinking and outlined by the focus on questions during brainstorming are often successful here, especially in maintaining stakeholder engagement (Redante et al., 2019).

### 8.2.1 Reasoning

Reasoning is something that is natural to design. When proposing solutions, designers look at the current situation and the design goals. They then reason to deduce what is the best way to create a workable solution. There are three universally recognized ways of reasoning:

- *Deductive reasoning*, where rules based on general principles can be applied and lead to a guaranteed solution. Deductive reasoning is often the practice, for example, in engineering design.
- *Inductive reasoning*, where specific examples are used as evidence of what works and how it can be used to provide the most likely solution. It is common in the solution-based approach.
- *Abductive reasoning*, which is primarily taking a best guess dealing with imprecise information, often using the what if question. It differs from inductive reasoning as the emphasis is more on cause-and-effect relationships rather than general rules. It is common in the value-based approach.

### 8.2.2 Brainstorming

Innovative solutions are often needed where information is not sufficiently precise to deduce a formally developed precise and correct solution. It is in those situations that designers increasingly rely on inductive or abductive reasoning to develop what might be called ideas that can be tried but does not guarantee a correct and precise outcome. Brainstorming is increasingly used in such reasoning.

The outcome must be a product or service that satisfies several criteria. The questions to be addressed in brainstorming are as follows:

- What service are needed to deliver value to customers?
- What are the data needs to deliver the service?
- What technologies are needed to deliver the service?
- How can the service be put into practice by integrating the service into the infrastructure?
- Can the service be justified on a cost-benefit basis?

These questions cannot be treated as independent problems. A service for, example, can only be proposed if it can be supported by technology. These questions thus cannot be done in sequence as there are strong links between them. They must develop, at the same time finding the best ways to combine ideas from more than one discipline.

**Fig. 8.1** A brainstorming environment

Brainstorming session includes the following:
- Several people raising ideas and discussing them.
- Visualizations to focus the discussion.
- Questions such as "What If we do something?"
- Post-it notes to record ideas.
- People standing while discussing ideas. Standing often leads to ideas emerging quicker (Fig. 8.1).



Brainstorming and reasoning are combined to search out ideas on ways to address issues. It takes place at all stages of any design. While brainstorming, people often stand in front of a model. They look at what is in front of them and make comments or suggestions on what can be done. They can suggest changes at some part of the diagram and then reason whether the suggestion leads to better solution by asking questions on all aspects at the same time.

There is no standard process for the sequence in which comments or questions are made in brainstorming. Brainstorming is thus often seen as totally freeform, but nevertheless, eventually it must deliver a useful outcome—in our case add value to stakeholders. No doubt everyone has seen people sit around discussing whatever comes up and producing endless post-it notes, diagrams, and reports, without any outcome. However, it is necessary to make sure that brainstorming does produce an outcome rather than simply a lot of recorded discussion. Hence there is often a facilitator who guides a brainstorming session, guiding the reasoning, recoding decisions as they are made to guide the session to an agreed outcome. The facilitator should also guide the reasoning used but not focus on the problem. For example, pose a goal like "What are the alternatives here?" rather than "What if we start a delivery service?" The facilitator thus should not take a leading role by suggesting solutions but guide the brainstorming team to them.

Brainstorming teams should be multidisciplinary. Such teams have a wide range of expertise not only in the context that we are working on but in other contexts. Having team members with expertise raises the probability of someone seeing how a solution used elsewhere can be adapted to the current context—following an inductive reasoning approach. Teams should also include team members familiar with different kinds of reasoning. Participants in such brainstorming sessions should be aware of all the possible choices with data and technology—hence cross-disciplinary teams are often an advantage.

Depending on the issues, the design team can proceed in several ways. One is to brainstorm ideas to suggest solutions by applying different reasoning. We can then rank the ideas and develop more details of those seen more relevant. Facilitators must ensure that people connect and work together. In increasingly remote work, it is necessary to develop technology platforms to support their work. The question then is how to combine brainstorming with reasoning so that a team can deliver a solution. Two commonly used ways are described here - the solution focused approach and the value based approach.

## 8.3 Solution-Focused Approach—Starting with a Solution Idea

Here solution generation starts with a suggestion like "Here is something that has been used by someone—I am sure that we can apply it here" or "There is a new technology that is emerging—we can apply it here to get some additional value." Brainstorming most often uses inductive reasoning trying to find ways to adapt a solution by using knowledge from earlier experiences, perhaps adopting the solution incrementally while learning to adapt in small steps at a time.

Many start-ups, for example, often start with a potential product, often emerging from research or services and look for ways to deliver value to others. Teams here use abductive reasoning to see älternative ways to deliver value to stakeholders. Many successful businesses have started that way. McDonald's, discussed earlier, started with a hamburger shop that focused on busy travellers and gradually, through experimentation in their product, grew from a local shop in California to a global enterprise.

- Start with a possible solution and how to create value in an existing system.
- Reason of how an idea can be adapted to satisfy stakeholder value.
- Continually adapt to a growing persona as their needs change.

The reasoning here is not followed in any sequence, but chosen by the design team as brainstorming proceeds. A common approach is to use mind mapping.

In ◘ Fig. 8.2, the design team might start with a solution that is often a newly invented technology. Principally, based on design thinking, the design team follows a line of questioning that starts with "where can we use it" and later with "what if" suggestions for change to address emerging needs. We have a look at stakeholder value judgements and may then look at ways to move from a negative to a positive judgement.

◘ Figure 8.2 gives one scenario for brainstorming using *mind maps*. Here brainstorming starts with asking questions about data or technology. What if we use a particular technology at some point of the process? Will it change stakeholder evaluation? While evaluating options, you might sketch a conceptual model in front of you to evaluate the effects.

- What if we do something?
- How does it affect stakeholders?
- How does it affect the process or journey map?

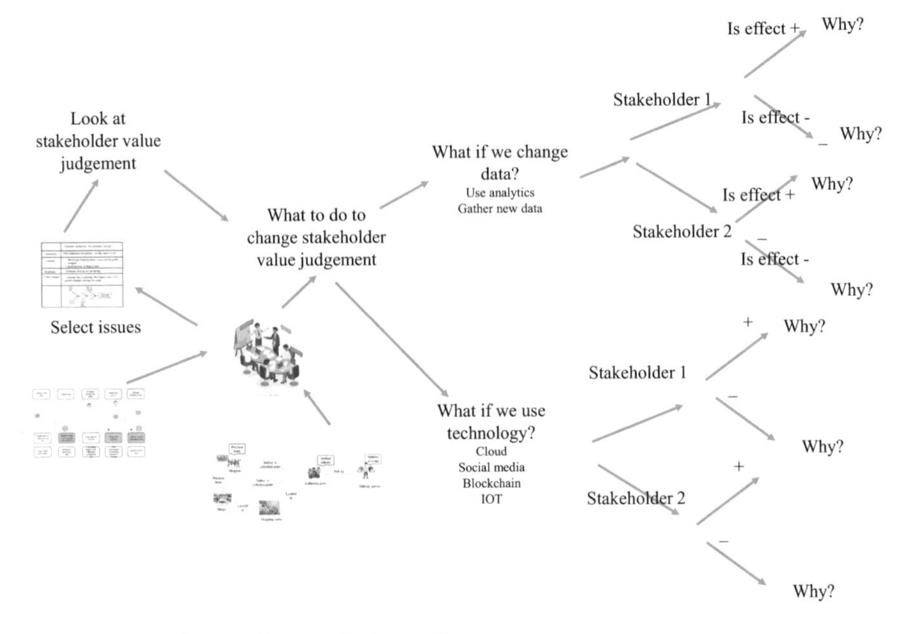


Fig. 8.2 A mind map for open brainstorming

### 8.3.1 Making an Idea Work

One important part of brainstorming is how to make an idea work. Again, it is possible to start with questions. What are the questions to ask? These become now extremely focused. Who is involved at each part and so on?

**Questions to Ask Include**

Do we need to change the existing process?
Do we need new activities?
How to improve journey map?
Where to get the data needed for each touchpoint?
What technology support is needed?
What services should be provided?

### 8.3.2 Where to Start?

Designers often start by looking at models and suggesting changes. One may be to suggest changes to journey map and from then identify the data needed in the changed process. It also often occurs that people in the brainstorming have their experiences with statements like "when I was in some location what we did was something ………." From a knowledge perspective, what we are doing is bringing in tacit knowledge into the team by bringing in people that can contribute to the solution through their expert or specialized knowledge—hence the importance of having multi-disciplinary teams. Such suggestions are spontaneous in most cases where people see emerging patterns that trigger patterns from their earlier experience or knowledge.

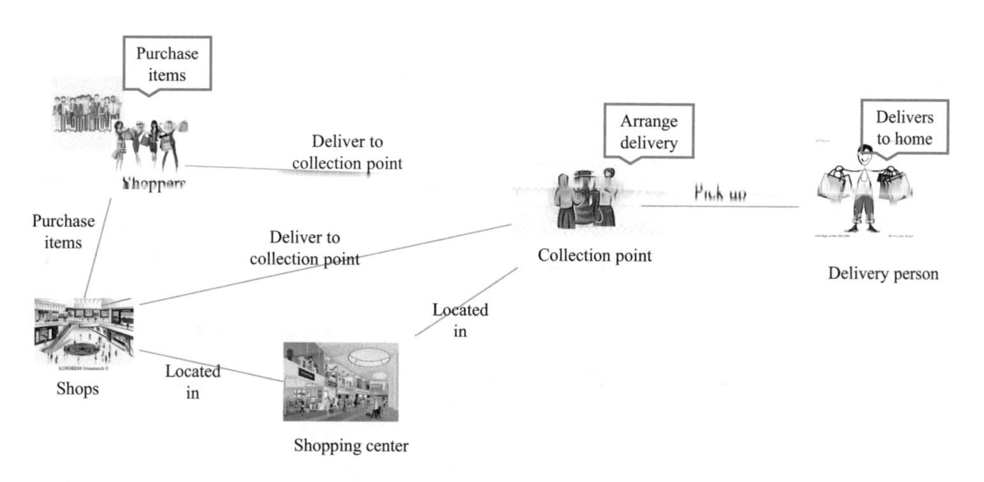


**Fig. 8.3** Customer-oriented solution focusing on service

The suggestion may be to continue with the idea to "Provide home deliveries." Making home deliveries may not be a new idea globally, but it is an innovation for a shopping centre that does not currently provide delivery services.

### 8.3.3 Sketching a Solution

How to start? What systems are needed and how will the process work? Who is involved? How will their work change?

A common way is to start with a sketch, for example, like that shown in Fig. 8.3. A sketch like that provides a focus for discussion on what will happen at each point of the process. It shows what the solution will look like; here shoppers will need to deliver their purchase to a collection point. The collection point will then organize for the customer; they buy it in the shop and then see it at home. There are of course alternate solutions where the delivery person picks up purchases at the shop.

Journey maps are often useful. Here we have journey maps in front of us—we can then ask how to improve journey maps. So, we can ask questions like what can be improved at each part of the journey. For example, why can't the shopkeeper arrange the delivery to the collection point? To answer such questions, we need to see what happens at each touchpoint. Is there a lack of needed data, or are decision-making tools needed?

## 8.4 Value-Focused Approach—Starting by Identifying Needs

The alternate approach is to structure the process where we still get the benefit of open ideas, but do so in a systematic manner. The solution generation usually starts with "We have now identified what we need to do. How do we do it?" These needs have been identified in earlier chapters and recorded in an issue note, such as that shown in Fig. 8.4.

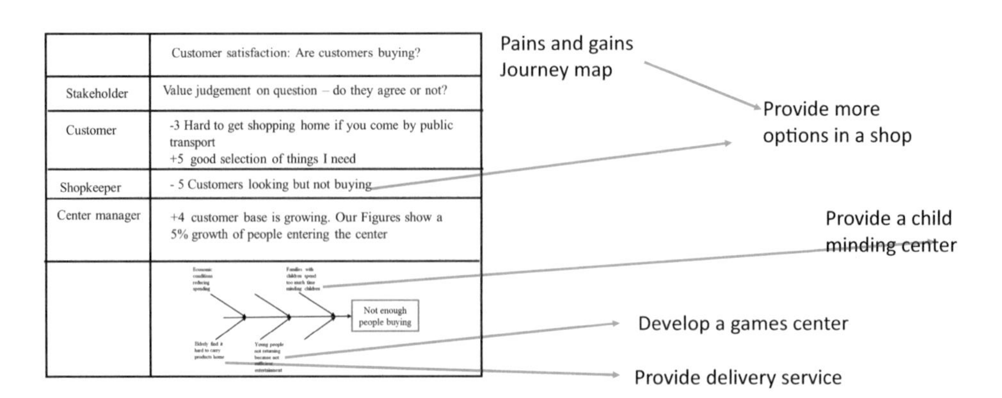


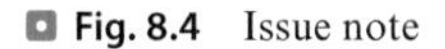
Fig. 8.4 Issue note

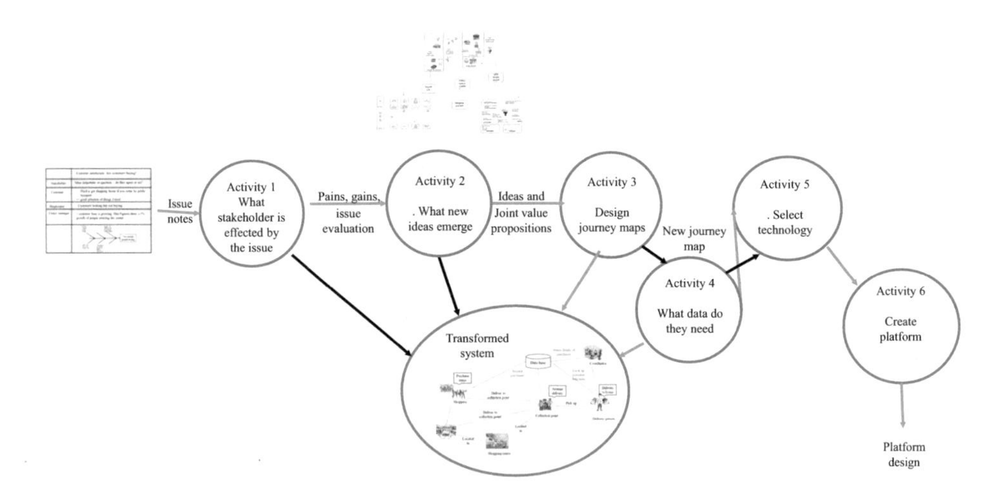


Fig. 8.5 Brainstorming towards an outcome

Design starts with identifying ways to address the problems that cause issues. In the context of a shopping centre, some people may see parking as important, others see helping with purchases and deliveries.

Figure 8.5 shows the kind of activities, six in Fig. 8.5, that take place in developing/brainstorming a solution. Each of these themselves have options and call for some brainstorming to identify possibilities and reasoning to select the best options. They need not necessarily be followed in sequence, but the double arrows between the transformation team imply that at any time a previous activity may be reactivated depending on outcomes in later activities.

### 8.4.1 Activity 1—Listing Stakeholder Concerns

One early step is to bring together all stakeholder's issues and concerns. These have been collected in early interviews described in ▶ Chap. 3, and later summarized in

Table 8.1 Stakeholder issues

| Stakeholder | Pains and gains | Issue evaluations |
|---|---|---|
| Customer | Not finding what they need<br>Difficult to find parking space<br>Buying what they need and not being able to carry it | Hard to get shopping home |
| Shopkeeper | Customers not buying | Most people looking, not buying |
| Centre manager | Lower number of people visiting Centre | Should be more sales |

▶ Chap. 7. The driving factors here are stakeholder value, especially their pains and gains, and their evaluation of the importance of the issue. So, it may be a good idea to list these in one place or document as shown in Table 8.1.

Questions to ask here may be:

- What stakeholders are affected and how is stakeholder value generated?
- What data they need to carry out these tasks?
- What technology is needed to both deliver the data and support the process?

### 8.4.2 Activity 2—Emerging Ideas Through Brainstorming

We use the information gathered earlier in Activity 1 to ask WHAT IF questions during brainstorming. So, four ideas shown in Table 8.2 may emerge. Table 8.2 illustrates in a limited way the argumentation process that starts with a WHAT IF question and follows with HOW to realize the idea?

It is possible to take each idea and develop a solution for the idea. Then compare each solution and select what is considered the most suitable one. Of course, we cannot go on with each of these ideas. One guideline is to develop SMART ideas—**s**pecific, **m**easurable, **a**chievable, **r**ealistic, and **t**imely. It is often possible to rule out some ideas early—for example, expanding the parking system (Idea 1) is not realistic—it is just too costly. We can probably rule out others, like option 3 as delivery is usually the responsibility of individual business, and shopping mall management does not at this time want to set up a mall shopping website.

So, then we might look at each proposal and select one to analyse in more detail. So, in the case of the shopping mall, a decision may be made to focus on home deliveries (Option 2 in Table 8.2). Often this decision can take the form of a joint value proposition. The joint value proposition will state what will be done to increase values. The value preposition is that "we (the transformation team) will develop a delivery service to help customers get their shopping home, and shops to increase sales in shops."

The joint value preposition identifies what we will do. The next step is to decide how to do it and how the new system will work. Journey maps are a common tool used to do define how the new delivery system will work.

Table 8.2 Idea table for projects

| Idea | HOW to implement the idea | WHAT IS the impact on stakeholders |
|---|---|---|
| 1. WHAT IF we2. Expand the Car Park? | Provide more car spaces to shorten the distance for carrying purchased items to car | Customers finding it easier to find parking |
| 3. WHAT IF we provide home deliveries? | Provide a service to arrange delivery of sold items to the home | Customers potentially will buy more |
| 4. WHAT IF we develop web-based shopping? | Develop a web-based shopping option where customers place orders in the shop for delivery | Customers will not need to travel to shopping Centre as often<br>Shops that rely on people being there will lose business |
| 5. WHAT IF we provide a trolley service? | Provide a Centre-wide trolley service where purchases are placed in a trolley and can be wheeled to the car | Customers will need to navigate trolleys throughout the Centre and to the car park |



### 8.4.3 Activity 3—Brainstorming About Process

Next, the brainstorming should look at HOW to implement the idea. Journey maps are one important way to identify the data needed by people. Each touchpoint is an interaction between people where they both exchange knowledge and need knowledge to decide.

Journey maps are crucial to any transformation as they focus on explaining what process people will use to carry out their activities and show where value is still not realized. In many cases, brainstorming begins by identifying issues found in journey maps or with persona pains and gains.

Here design issues like the following can be considered.

Where is the greatest dissatisfaction in the current journey map? Maybe we can change their journey to eliminate points of dissatisfaction.

Maybe we should address the last touchpoint to see where awareness and research needed to make decisions can be simplified.

Look back at previous journey maps in ▶ Fig. 6.2 in ▶ Chap. 6 that identified where stakeholders found needed to be addressed. We can then develop a new journey map, which is shown in Fig. 8.6. Here we arrange for purchases to be delivered to the shopper's home.

What we have done is addressed those touchpoints that caused most pain or negative feeling as identified in journey map touchpoints. For example, we can capture the persona of people that use the service and focus our marketing to bring them to the centre. Thus, in Table 8.3, the shop assistant not only arranges delivery but also records future client needs. Or we can capture what people are buying and when and provide a delivery service that shopkeepers and their customers can use.

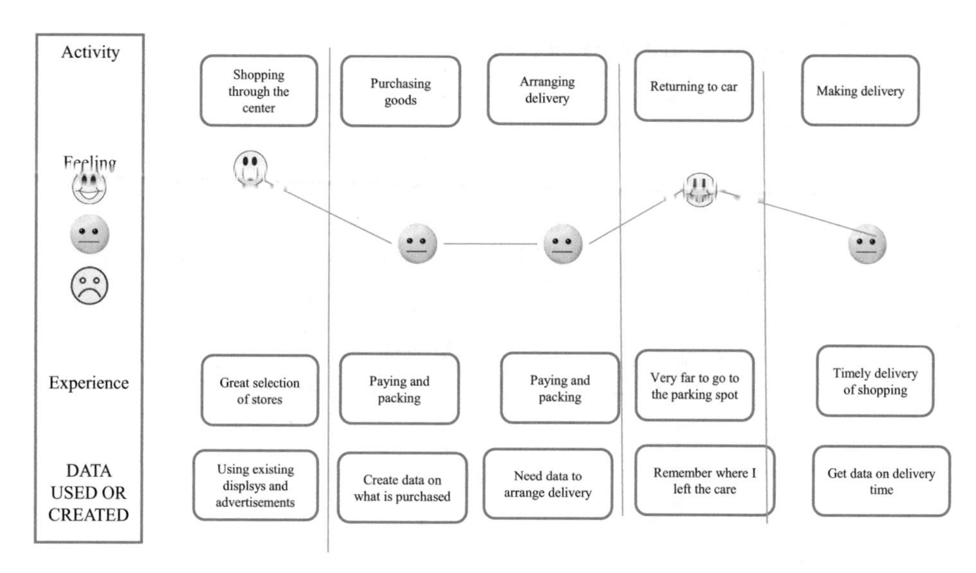

**Fig. 8.6** New journey map

**Table 8.3** New touchpoint

| Journey Map: Arrange delivery at time of buying in the shop | | | |
|---|---|---|---|
| **Touchpoint: Arrange delivery in store** | **Customer** | **Shop assistant** | **What are the data needs** |
| Awareness (finding what is possible) | Provides their needs to the shop assistant. | Arranges possible delivery times. At the same time records customer needs as potential later sales. | List of products on sale. |
| Options (making a choice) | Chooses the products that most meets current needs. | Discusses possibility of meeting future needs. | Evaluating how to use the product. |
| Research (what is the best option) | Think of most convenient delivery option. | Shop owner sets priorities for deliveries. | Making a choice. |
| Selection and action | Select a delivery time. | Arranges delivery at purchase time. | Delivery items and address. |

### 8.4.4 Recording the Transformation on the Rich Picture

Now we show the changes to be made in the transformation of the system. These are shown in Fig. 8.7 using the system model. These are shown by the oval shapes as follows:

- The addition of a new stakeholder, the delivery person.
- The introduction of a delivery service.

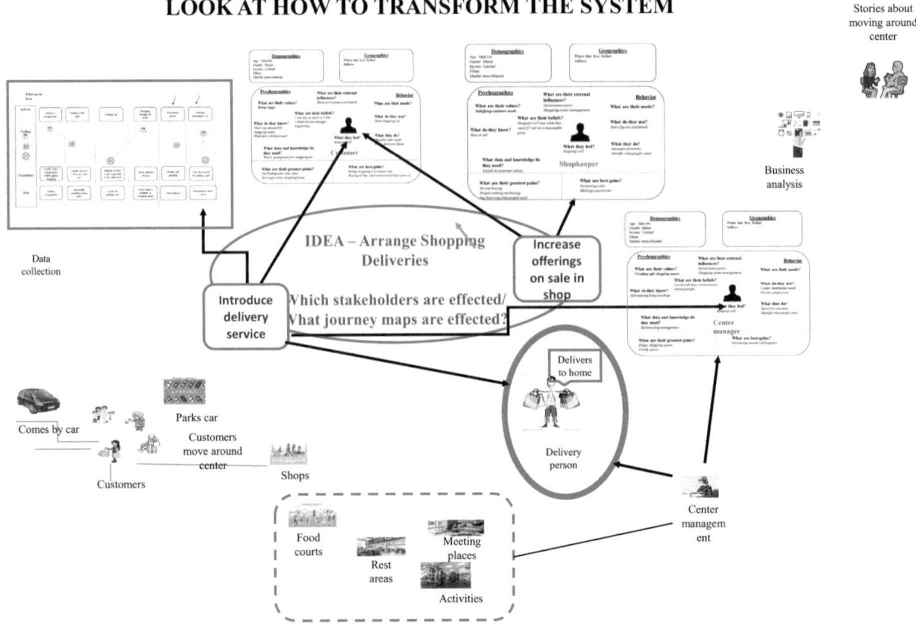


**Fig. 8.7** Transformation on rich picture

Now we begin to look at more detail, especially at new data that may be needed and technology to support the transformed system.

## 8.4.5 Activity 4—Brainstorming for Data Needs

In almost all cases, a new system will need new data. The data can be about clients, where to set up services, or emerging technologies. Data needs can be met by:

- Providing new data needed in proposed new stakeholder tasks.
- Providing additional data to existing people in the system who are looking for new services.
- Making existing data more widely available and make sure that data to both get it to the right place but also to use it to make decisions.
- Analysing existing data to provide better insights.
- A mix of the above.

When discussing data, it is sometimes necessary to make a distinction between data, information, and knowledge. Such distinction was described earlier in ▸ Chap. 5 and is shown following.

- *Data* is elementary facts captured at the time of a transaction—for example, a sale.
- *Information* is where data is organized in ways that show some combination or trends—for example, grouping sales by store or item to produce summaries or trend.
- *Knowledge* is the interpretation into decisions on what to do.

Modern thinking is that businesses should become smarter is emerging. Becoming smarter usually means analysing data to create knowledge that allow stakeholders to make better decisions. Thus, a transformed process must not only capture new data but also analyse it to create information for stakeholders, who then use it to create new knowledge mainly of how we can use the information to make better decisions (◘ Table 8.4).

It is not only getting the right data but making sure it is carefully analysed and presented to decision-makers. For example, the existing journey map shown in ► Chap. 6 in ► Fig. 6.2. It showed shopper dissatisfaction with having to carry large purchases home.

### 8.4.6 Activity 4a—WHAT Information Is Needed

Data analytics are used to process data into information and knowledge needed by stakeholders to make decisions. There are of course many options to providing information by arranging data, in different ways and using different visual presentations (◘ Table 8.5).

◘ **Table 8.4** Brainstorming what is needed to make idea work

| Idea | WHAT stakeholders need to do | WHAT data is needed | WHY is the data needed? | HOW can I get the data? |
|---|---|---|---|---|
| Provide home deliveries | Customer calls recommended delivery person. Shopkeeper hands to delivery person. | Contacts of people who deliver | To find someone to deliver a customer purchase | Lists of delivery companies |

◘ **Table 8.5** Smart data options

| WHAT information is needed? | HOW do we get it? |
|---|---|
| Collecting and identifying big data | The vast amount of data now available through technology. Searching data using relevant keywords can provide solution ideas |
| Using business analytics to make better decisions | Rearrange data to show important relationships |
| Provide ways to evaluate option | A table showing problems in supply chains can guide designers in identifying a supply chain problem |
| Situation visualization | Visualizations such as maps, graphs, and models that promote better understanding of a problem. Showing activities like sales on a map is an example |
| Making data available to wider community | Developing platforms for distributing data and analysis results to people |

Increasingly now, technology can provide information, but with advances in artificial intelligence (AI), it can also provide knowledge. One possible example is in monitoring by identifying patterns and then suggesting solutions.

### 8.4.7 Activity 5—Getting Value from Technology

Technology plays an increasingly important role both in managing access to knowledge and in creating platforms for users. There are now many options available. Each of these of which adds business value. Some are shown in ◘ Table 8.6.

## 8.5 Choosing Technologies

The choice of technology is now staggering. The possibilities to get value from effective use of technology is almost limitless. But it requires the right choice. The choice often depends on whether some issue in a journey map is addressed or whether there is a system-wide adoption.



### 8.5.1 Potential and Available Technologies

There are now numerous technology options that can be used in design. Each contributes in different ways to solutions. Some potential contributions are shown in ◘ Table 8.7.

One technology that is now increasingly mentioned is artificial intelligence.

◘ **Table 8.6** Business value generated by technology

| WHAT Business Value will be created | Possible Benefits to business value through Technology |
|---|---|
| Maintaining and growing a customer base through improved customer satisfaction | Maintaining a loyalty program.<br>Reducing delivery times<br>Profiling with big data as a key activity to identify markets and trends<br>Reducing costs of purchasing |
| Maintaining a strong internal environment | Better coordination through platform for sharing knowledge about customers with internal staff<br>Can assist in recruitment<br>Making education courses available to a wider audience |
| Being relevant | New product development<br>Getting access to wider information and expertise |
| Working with partners | Brings partners together into a common space |

Table 8.7 Technology contribution to business value

| Technology | Potential contribution to agility and business model innovation |
|---|---|
| Cloud | Providing platforms for sharing knowledge and collaboration. Often cloud platforms can lead to new business models as they can provide a wide set of services globally. Cloud technologies are increasingly used to support connectivity between all the people involved in the business process |
| Social media | Communications and knowledge about personas. Use social media for contact between business and its clients. Use knowledge in social media to create new services |
| Mobile devices | Mobile phones and other communication devices which, when combined with cloud and social media, result in greater mobility |
| Mobility | Add to possible disruption as existing business models can now become global and hence commercially viable |
| Artificial intelligence (AI) | Learns from previous experience and suggests what to do at a touchpoint |
| Blockchain | To ensure the quality of any inputs delivered to each supply chain stage. |
| Internet of things and sensor devices | Technical devices that monitor data in an organization. Can be used to collect information collected earlier by manual means or not available |

### 8.5.2 Choosing to Transform Journey Maps

The most straightforward choice is where changes are made to stakeholder journey maps (Table 8.8).

The remainder of this chapter outlines some possibilities in the choice of technology for designers.

## 8.6 Transforming Systems

The trend now is to use technology not only to simplify work digitization but also to improve all parts of the business model—in fact transforming the way business is done. Often the choice of technology depends on the industry or the type of activity. There are now many trends emerging in business. Designers should be familiar with these as they may be useful in their problems. Common examples include the following:

- Sharing resources in the sharing economy.
- Supply chains for product development and delivery.
- Getting expertise to different parts of any process.
- Working with partners.

**Table 8.8** Making the choice

| Stake-holder | Task | Data | Technology (Web, iPhone, Big Data, IoT, workflow, etc.) | How will the stakeholder use it |
|---|---|---|---|---|
| Customer | Requesting delivery | What is to be delivered<br>Address of where it is to be delivered | Mobile phone | Monitor and discuss details with delivery person |

### 8.6.1 Technologies for the Sharing Economy

Earlier chapter of the book described some examples of sharing economy, especially Airbnb and Uber. The technology here are platforms where sellers advertise their services while buyers access the services. Technology has been practiced for many years now as exemplifies by online trade and online shopping. Uber and Airbnb are new examples of how technology can be used to create new ways of doing business.

Important technology here are platforms to provide services and bring suppliers and clients together.



### 8.6.2 Global Supply Chains

The globalization of business increasingly leads to several businesses working on different parts. Virtually any industry has always been part of supply chains. Food production is one of the most typical. Motor vehicle production is another. Each part of production process brings in its own expertise and includes the following:
- Making contractual arrangements with other businesses, especially those providing inputs and purchasing the outputs.
- Arranging the delivery of any materials or expertise needed and ensuring that correct goods are delivered.
- Making design and production decisions.
- Arranging the production of products and services.
- Maintaining storage for parts.
- Marketing, distribution, and delivery of the products and services.
- Increasingly, each business in the supply chain is required to maintain a low carbon footprint.

Supply chains like that in Fig. 8.8 use technology at all stages. One important goal of any technology in the supply chain is to improve connectivity between all parts of the process. Another is traceability to determine that the source of all products at any step meets requirements. Traceability is particularly important in food chains as many people like to know the source of the food they eat.

## 8.7 Platform Design

The final part of the design is to broadly specify the platform through which services will be delivered to stakeholders. The platform must support several stakeholders, each requiring some different services, but often using the same data. Journey map touchpoints give an indication of service to be provided. For deliveries, the main touchpoints can be recording a sale together with a delivery request. Another touchpoint is the actual delivery, which can include the customer and delivery person.

Figure 8.9 is a diagram illustrating a platform. Each stakeholder, as shown in Fig. 8.9, is supported with a different interface and devices. An application program must be developed for each activity in the system.

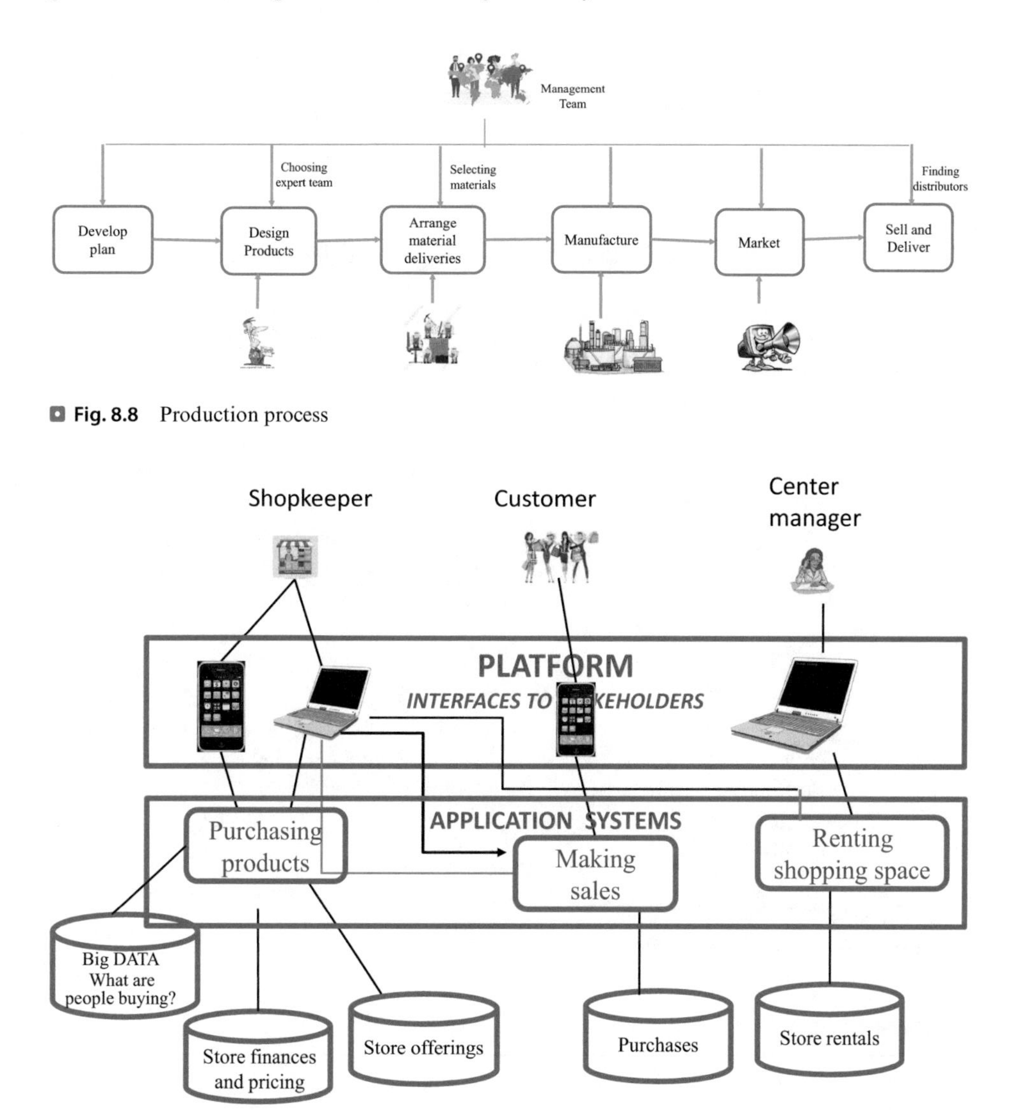


Fig. 8.8 Production process



Fig. 8.9 Designing a platform

Apart from satisfying current needs, the platforms must satisfy criteria that can support platform evoltion as for example:

- Provide for growth not only in number of customers but also range of services. Amazon, for example, has diversified into many areas, Amazon started selling books, but now has extended to virtually sell all kinds of products.
- Allow customization—Apple allows apps to be placed on its platform, as does YouTube.
- Emergence in new directions.
- Learning self-organization.
- Support connectivity as for example, LinkedIn.

**Summary**
This chapter described ways to create solutions that satisfy stakeholder values. It suggested different ways to generate solutions but focused on a value-based approach that starts by identifying needs.



**Exercise**

Develop some ideas on ways to address the issues found important in ▶ Chap. 7.

You can organize into teams and begin with proposing ideas that address issues identified in ▶ Chap. 7. Brainstorm some solutions and post them as post-it notes. Follow the activities in ▶ Sect. 8.4, recording the outcomes.

The solution for each case should include the following:

- A table like ◘ Table 8.1 for the different stakeholders in the case.
- A table like ◘ Table 8.2 to list the ideas.
- The argumentation to select an idea to develop in more detail.
- A new journey map for the selected idea.
- A table like ◘ Table 8.3 for your data needs and.
- A table like ◘ Table 8.8 for suggested technologies.

You should use the issues you have identified in ▶ Chap. 7 to generate your ideas. For restaurants, for example, you might for example focus on a problem introduced in ▶ Chap. 5 on providing meals for vegetarians. One issue may be to provide a range of dishes to attract a wide range of vegetarians. How do you find such a range of dishes? Would technologies like big data help? Or maybe you can collect feedback from your customers. And how will you know that you made the right choices.

## References

An, D., Song, Y., & Carr, M. (2016). A comparison of two models of creativity: Divergent thinking and creative expert performance. *Personality and Individual Differences, 90*, 78–84.

Dorst, K., Kaldor, L., Klippa, L., & Watson, R. (2016). *Designing for the common good.* BIS Publishers.

García-García, C., Chulvi, V., & Royo, M. (2017). Knowledge generation for enhancing design creativity through co-creative Virtual Learning Communities. *Thinking Skills and Creativity, 24*, 12–19.
Redante, R. C., deMedeiros, J. F., Vidor, G., Cruz, C. M. L., & Rebeiro, D. (2019). Creative Approaches and green product development: Using design thinking to promote stakeholders' engagement. *Sustainable Production and Consumption, 19*, 247–256.
Taura, T., & Nagai, Y. (2017). Creativity in Innovation Design: the roles of intuition, synthesis, and hypothesis. *International Journal of Design Creativity and Innovation, 5*(3–4), 131–148.

# The Wider Context

Contents

# Managing in Disruptive Environments

## Contents




I. T. Hawryszkiewycz, *Transforming Organizations in Disruptive Environments*,
https://doi.org/10.1007/978-981-16-1453-8_9

In this chapter, we show how solutions can deliver values in disruptive environments. It describes the additional best practices needed to achieve the resilience and agility to operate in disruptive environments.

**Learning Objectives**

- What are disruptions?
- Agility and resilience
- How to develop resilient systems?
- What causes disruptions?
- Mitigation, response, recovery
- Transforming to make organizations resilient

## 9.1 Introduction

"Disruption" is now a commonly used term in industry and society. However, disruption is not a word that has a rigid definition and is frequently used in informal ways to describe something unexpected that must be addressed. Disruption can be minor such as a customer coming back to return a faulty product to a store. It can also be major natural disaster such as a flood, major or economic and health crises such as COVID-19. Responses to each disruption may be different. This chapter first defines the term disruption and its difference from crisis, and then describes how to manage to minimize any damage caused by a disruption.



This chapter begins by describing different kinds of disruption. It also identifies the need for organizations to be resilient against disruption by defining ways to minimize the impact of disruption on stakeholder values. It develops additional best practices to use in the methods described in earlier chapters to ensure transformation uses best practices to deal with disruption.

This chapter then concludes by describing examples of different kinds of disruptions and the ways they have been addressed.

## 9.2 What Are Disruptions?

There are many definitions or at least descriptions of disruption. Natural disasters such as bushfires, floods, earthquakes, and landslides are obvious as causes of disruption. There are also economic disruptions such as recessions, as well as epidemics or pandemics that result in threats to health. There can also be internal business disruptions such as a new leadership team, or a decision to develop a new service, or an emerging competitor. Some definitions of disruption are given below.

- Unexpected events which interrupt a business activity or process with long-term effects.
- Actions that prevent something, especially a system, process, or event, from continuing as usual or as expected.
- Actions to completely change the traditional way that an industry or market operates by using new methods or technology.

The major causes of disruption are now seen as:

- Natural disasters such as bushfires, earthquakes, or floods.
- Manmade disasters such as war or rioting.
- Economic disasters such as depression.
- Epidemics and pandemics, and.
- Disruptive innovation, which can affect how whole industries work.

### 9.2.1 Disruptions and Crises

It is also important to distinguish between disruptions and crises. A crisis happens when an organization is unable to respond to a disruption. A bushfire is a disruption, whereas a crisis would be where there are too few fire trucks to put out the fire. COVID-19 is a global disaster that caused a few crises, as for example, the lack of ventilators and hospital beds that led to otherwise preventable deaths in some countries.

One other difference between a crisis and a disaster is that you can often assign responsibility for a crisis, but not for a disaster. For example, a bushfire can be caused by a lightning strike. No one can be blamed for a lightning strike that caused a bushfire. But blame can be assigned if there are too few fire trucks and the bushfire gets out of control. You cannot blame anyone for a pandemic, but you can blame administrators for a lack of hospital beds. The lack of hospital beds can become a crisis.

### 9.2.2 Types of Disruption

Disruptions can be described in several ways—by their scale and size, their scope, and how quickly they emerge. Some examples are shown in ◘ Table 9.1.

◘ Table 9.1 is organized so that as you go down further in the table, the disruptions are greater, and response becomes more urgent. The later the response to a disruption, the greater the damage. The damage by a fire left to burn is greater than when you begin to put the fire out at the start. ◘ Table 9.1 does not include climate change that leads to slow and emerging disruptions, even possibly disasters, sometimes leading to crisis situations. Climate change is considered at length in ▶ Chap. 10.

One of the difficult disruptions to understand is disruptive innovation, as, to many people, innovation is seen as a positive activity. It can, however, be disruptive to those that prefer continuity in their work.

### 9.2.3 What Is Disruptive Innovation?

The term disruptive innovation is often used to describe the impact of innovations in technology in the way things are done. Disruptive innovation has been happening since the start of the Internet, which has changed the most elementary ways of

Table 9.1 Classifying disruption

| Type of disruption | Example | Scope—what is impacted | Response | How quickly does the disruption emerge |
|---|---|---|---|---|
| Everyday business | Emerging competitor | A business | Assembling teams to develop strategies respond to new competitor | Usually, gradual. Can sometimes be predicted |
| Disruptive innovation | Uber, Airbnb | Industry as, for example, the taxi industry | Develop policies to change the way an industry works | Usually, gradual. Impact can be predicted |
| Natural disaster | Bushfire, earthquake, flood | Local area, city, businesses in the area | Preservation of life and property around disaster | In most cases, rapidly emerging, requiring urgent response |
| Economic financial | Trade war, recession | Large, global | Community financial well-being through financial mechanisms | Emerging onset as businesses fail |
| Epidemic, pandemic | Ebola, COVID-19 | Large, global | Preserving community health through health services | Quick onset if disease is easily transmitted |



doing business, such as e-mail instead of sending letters. It has provided the ability to connect to customers through interactive interfaces, profiling customers, to add to business value of retaining customers.

Disruptive innovation is often characterized by the emergence of new technologies that lead to new consumer habits, and new government policies. Such change impacts on every business as it encounters new competitors, government decisions, and adopting new technologies in their business activities.

It is, however, not the first-time technology has changed the way we live. Disruptive innovation in business has been a constant phenomenon. The development of rail travel in the 1800s led to the growth of the leisure travel industry, which, to many, was a positive disruption to everyday life. The travel industry was then given more impetus with the growth of air travel.

Disruptive innovation changes what we do. In that case, Uber provides a different way to get a service to go somewhere. For current taxi businesses, it is a new competitor. Mobile phones can also be a technological disruption that has had wide impact, as now we can call anyone from anywhere, and do business from anywhere.

## 9.3 Managing in Disruptive Environments

Irrespective of the kind of disruption, organizations are developing processes to manage in disruptive environments. Such processes include three parts:

- *Mitigation*: anticipating the disruption and taking precautions to minimize it effect
- *Response*: minimize the damage once an event occurs, and
- *Recovery*: to get back to normal after the damage has occurred.

The innovation required in all three phases are different. Generally, mitigation and recovery call for a value-based approach to prevent or reduce damage, whereas response calls for quick solution-based action to reduce damage. Organizations must support people living through disruptions, whether natural or otherwise. The terminology now commonly used is to develop resilience to cope with disruption.

## 9.4 Developing Resilience

Resilience and agility are becoming increasingly important topics in business and the community as now illustrated by the COVID-19 pandemic, where responses must be quickly developed to limit virus spread. Resilience to climate change, natural disasters, such as earthquakes, forest fires, and floods, is also seen as necessary.

Resilience is the process of *adapting well in the face of adversity, trauma, tragedy, threats, or significant sources of stress*—such as family and relationship problems, serious health problems, or workplace and financial stressors. It means "*bouncing back*" *from difficult experiences*. Agility, on the other hand, is the ability to quickly adapt to a disruption—for example, how to quickly adopt technologies to add value to their business or bring resources quickly to fight a bushfire.

In this chapter, we continue the approach where transformation uses best practices to identify issues to be addressed. We do this this by adding best practices for resilience and agility. The question then is, what are the best practices for resilience and agility? Figure 9.1 provides a framework for the best practices to be added to those described earlier in ▶ Chap. 4.

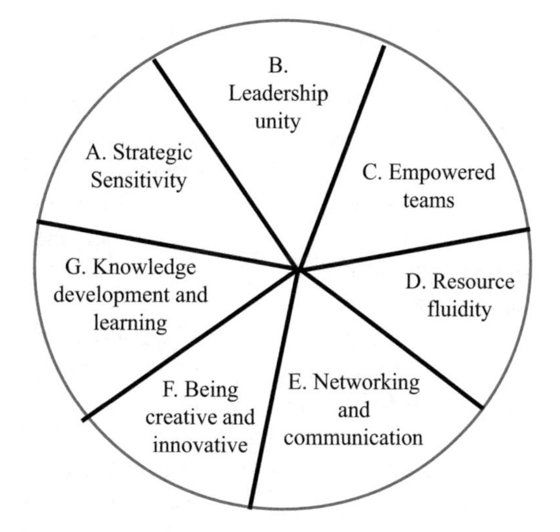


**Fig. 9.1** Resilience framework

- *Strategic sensitivity* where organizations continually scan their environment to identify potential disruptions. It is particularly relevant to anticipate any disruption from technology innovation.
- *Leadership unity* for organizations to respond in one direction, often to bring different agencies together under unified direction.
- *Empowered teams* to make decisions on the spot and not through some bureaucratic process.
- *Resource Fluidity* or the ability to move resources quickly to the location of the disruption.
- *Being creative and innovative* continually in any response.
- *Networking and Communication,* to support teamwork and collaboration.
- *Knowledge Development and Learning* especially in ways that learn from previous disruptions to prepare for the next disruption.

### 9.4.1 Mitigative Actions



Mitigation is based on the prediction of what can happen. Mitigation often follows a value-centred approach described in ▶ Chap. 8. What will be the damage caused by a bushfire? How many people will suffer health problems if a health issue is not controlled? How much income will be lost if a customer base is reduced?

What actions can be taken now to anticipate the disruption and reduce the damage? In general, the quicker the response the less the damage. For example, what are the best practices to be followed now to protect the impact on property that can result from a fire?

The focus here is to develop knowledge on potential disruptions and use strategic sensitivity to identify potential developments that can impact business value. It is also training response teams and developing team leadership.

### 9.4.2 Responding to Disruption

In many cases, disruption starts with some event. This event can be sudden, like an earthquake. It can also be an emerging event, for example, a perceived emerging economic crisis. Or it can be an emerging epidemic, such as Ebola in Africa and, more recently, the COVID-19 pandemic. These events affect some values more than others. For example, in a pandemic, health becomes a significant value to individuals; in a fire, safety and property become more significant; in a business, customer retention becomes more important.

Once a disruption is recognized, a response usually follows. In most cases, the initial response is to create a community or teams with the necessary skills to reduce the damage caused by the disruption. In a forest fire, the first response is often the arrival of a fire unit. Depending on the scale of a disruption, there may be more than one agency responding—for example, police to direct traffic around the fire, ambulances to deal with the injured, and agencies to house people displaced by the fire.

Response must be quick and calls for a solution-centred approach described in ▶ Chap. 8. If there is a fire, then a quick solution is to call fire units, who have the knowledge of how to minimize any ensuing damage. In health epidemics, the threatened value is health, and the immediate response was lockdowns to limit the spread of the disease.

Irrespective of scale, there are some generic ways to characterize processes followed when faced by disruption. These are shown in ◘ Fig. 9.2 and described in ◘ Table 9.2.

◘ Figure 9.2 shows the steps followed in most cases to address a disruption.

- First evaluate the impact of disruption on stakeholders' values. For example, businesses responding to competitors need knowledge about how customers adopt new services, and how to find the expertise to counter competitor offerings. In global disruptions, such as pandemics, data is needed on ways to address causes, the spread, and ways to minimize disease spread.
- Set up the networking and communication systems to enable teams to work in innovative ways. Communication must also be set up to inform affected commu-

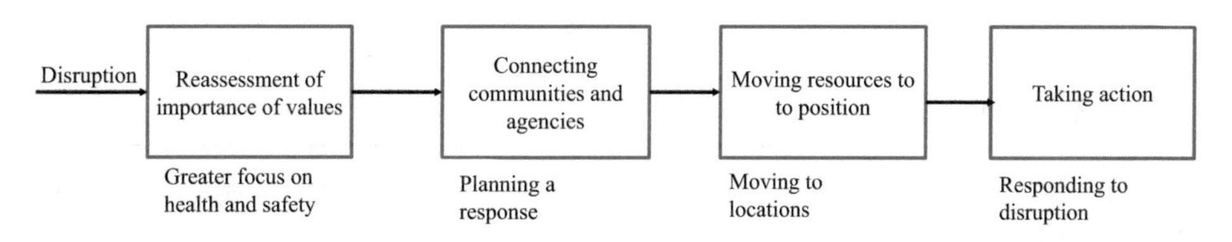


◘ **Fig. 9.2** A generic response model

◘ **Table 9.2** Activities in responding to disruption

| Response action | What needs to be done |
|---|---|
| Reassessment of values | Analysing value changes during a disruption focusing on risk to values from disruption. Health pandemics have always resulted in health rising as a value. In fact, that can become the dominant value |
| Connecting communities and response agencies | Identifies response agencies and links with communities affected by the disruption. In the case of health, communities focus on hospitals and medical services to reduce any negative impact on their health. In the case of a bushfire, they will need fire trucks, police, and possibly ambulance services. In case of a business facing a new competitor, a committee often develops a response using both internal and external data |
| Move and organize joint actions that call for leadership unity | Support of the mobility of resources both human and physical is often known as resource fluidity. Such fluidity will result in providing businesses and communities and teams with the resources as quickly as possible to respond |
| Take specific relevant action | Actions like providing medical care to individuals or putting out fires |

nities of any arising danger. In general, such communication is needed to develop knowledge of the progress of disruptions, and *resources needed to respond to any danger.*

- Identify resources needed to address the disruption, including leadership to create and empower response teams. For example, in pandemics, teams are empowered to test for infections and then trace infections to their source, and
- Support the processes taking the necessary action.

There is now evolving knowledge on ways to respond to different kinds of disruption. Processes are now established in many communities on ways to deal with potential disruptions. These are described later in the chapter. These processes are increasingly improving. We learn from earlier occurrences, identify issues, and continually improve response. Methods described in the second part of the book can be used here.

Response to disasters almost is a chapter of its own, but the methods described in ▶ Chap. 7 provide guidelines to identify the data needs and technologies to support response activities. Interested readers are also encouraged to read existing reports, especially to see how previous responses used the criteria described in ◘ Fig. 9.1 to develop resilience to disasters.



### 9.4.3 Are New Capabilities Needed?

The question now is how to enable resilience to be integrated into existing processes, and whether new processes and capabilities must be developed to do so.

These capabilities include collaborative response to a disaster, where all agencies unify their leadership to focus on the disaster. Leadership unity is thus one capability needed for effective response. The capabilities should lead to greater organizational resilience and agility.

## 9.5 Making Businesses Resilient

The term agility then refers to the ability of organizations and communities to *quickly change* what they do to respond to the disruption or minimize its undesirable effects. These can become additional best practices for businesses and cities to adopt and raise questions on issues to address to create resilience. ◘ Table 9.3 shows some best practices questions commonly used in business to address resilience. It also includes questions related to the environment—waste management, energy conservation, and emission standards.

## 9.6 Responding to Natural Disasters

Local disruptions are often caused by natural disasters, although terrorism in the past has also been a contributor. Examples of typical disruptions here are natural events such as bushfire, flood, or earthquake. They may also be events such as a terrorist attack, which may be local or have a global impact such as the 2001 attack on the Twin Towers in New York.

■ **Table 9.3** Best practices to mitigate effects of disruption

| **Business resilience themes** | **Questions to be asked to identify unsatisfied values** |
|---|---|
| Strategic sensitivity | Do you have a risk analysis program?<br>Do you continually scan the environment for threats or opportunities?<br>Are you collecting data on industry trends? |
| Leadership unity | Is it easy to change management responsibility?<br>Does your management team collaborate? |
| Empowered teams | Are your teams authorized to take action?<br>Are your teams empowered to respond to change?<br>Do your teams include people with relevant knowledge? |
| Resource fluidity | Can you quickly assemble a team to respond to an unexpected threat?<br>Can resources be quickly reassigned?<br>Do you have a stress testing activity?<br>Do you resource redundancy program? |
| Networking and communication | Do you provide ways for teams to communicate?<br>Is your response constrained by bureaucratic processes? |
| Being creative and innovative | Do you run brainstorming sessions that involve teams from different agencies?<br>Do you have ways to quickly evaluate ideas? |
| Knowledge development and learning | Do you capture data during a disaster? |

## 9.6.1 Local Community Mitigation

The major steps in response are to identify the values under threat. Property and lives are under threat from a bushfire, flood, or earthquake, whereas health is under threat in an epidemic or pandemic. Once threats are identified, capabilities needed to either eliminate the threat or reduce the damage must be developed.

What are the issues and capabilities needed? ■ Figure 9.1 provides some guidelines. For example, resource fluidity is needed to ensure resources can be moved quickly to the point of disturbance; another example, knowledge must be provided to communities on precautions needed to reduce any damage.

It is now common for authorities to provide guidelines and checklists of how to respond in case of a disaster and prevent damage to property. Provision of such guidelines is now seen as almost a duty of care in most jurisdictions, local, state, or national. There are many examples of guidelines or best practices advised to communities. Often these depend on the potential disruption, the community physical environment, and the kind of disruption.

**Bushfire Advice in Australia**

Advice provided by the Rural Fire Service in New South Wales, Australia. This advice is often provided through websites (see references) that provide the knowledge needed to prevent damage to homes or evacuation strategies. These guidelines make recommendations such as:

- Understanding bushfire warnings.
- Preparing homes to reduce damage.
- Understanding different alert levels.
- Maintaining communication systems.
- Understanding evacuation procedures.

Wildfires, as bushfires are sometimes known, are now increasingly happening in all parts of the world and jurisdictions in those parts provide advice on mitigation for their communities.

Advice provided by FEMA (Federal Emergency Management Service) through its website (see references}. This website recommends best practices to be followed to minimize risk of damage from fire, as for example:

- Ways to evaluate your risk.
- Keep combustible materials away from the house.
- Enclose eaves with wire mesh so that eaves do not gather in gutters.
- Reduce tree and vegetation around the home to minimize damage by fire, as for example making a 30 ft. safety net around the house and 15 ft. between tree canopies to stop fires jumping across trees.



There are many detailed suggestions in such advisory sites and readers working in these areas should access the sites for detailed descriptions. These and other sites show the importance of reducing the risk in situations where disruptions and resulting damage are frequent.

Any transformations should consider any risks of natural disasters in their area and follow any advice available for the area. Natural disasters include earthquakes, landslides, floods, and others. Communities are now warned of impending disasters and what to do as a response. Design of business systems in disaster prone areas should integrate such response into their design. Such advice is now available, as for example:

- Advice on floods in the United Kingdom can be found (see references).
- Conditions for landslides in Hong Kong in Choi and Cheung (2013).

### 9.6.2 Local Community Response

Disaster response requires leadership and community support. In many cases, especially in major outbreaks, calls for response from many jurisdictions and agencies. The response by agencies with expertise is needed to minimize the danger from the disaster. Generally, there are several agencies that must collaborate to address the disaster.

For example, response to a fire must include the fire department. But it also needs the police to control traffic flow, and ambulances in case of injury. Access to communication, power, and water is required as well. Collaboration between them must be quite close and almost continuous. Data plays an important part here—especially resources available for recovery, often availability of housing.

In many cases, response creates a control centre, which includes leadership of all agencies, which are involved in the response and monitor progress. The control centre would include senior people from all agencies to ensure coordination between personnel in the field.

### 9.6.3 Local Community Recovery

Once the emergency is over, the next step is recovery, to get back to normal as soon as possible. Depending on the extent of damage, recovery can include rebuilding, relocating people, and repairing any damage.

## 9.7 Global or National Disruption

As disruptions become larger, the response gets more specialized and detailed description becomes out of the scope of this book. Here the scope of activity and organizations involved are high. Disruption is across several communities and, in the case of pandemics, across many nations. Here leadership is needed to define rules for community behaviour and ways to mitigate against increasingly changing threats.

Data needs are now extended to sharing knowledge between communities, especially on ways different mitigation or response strategies have on the spread of damage. The following are two examples of national response.

### 9.7.1 Global Financial Crisis

The global financial crisis of 2007–2008 affected many nations.

The global financial crisis led to the loss of 8.8 million jobs, many home foreclosures, loss on stock markets and business bankruptcies. The impact of the global financial crisis has now been reviewed on many occasions as to the effectiveness of the response. A detailed analysis can be found on the Investopedia site (see references), which is the world's leading source of financial content on the web. The lessons learned can be found on:

Ultimately the outcome is a better understanding of the causes of the crisis and the evolution of best practices to minimize a recurrence. They included financial recommendations such as:

Stress testing of large financial institutions, which were too large to allow to fail.

Recommendations on ways to reduce risk and hasty lending.

Any analysis and resulting recommendations require expert financial knowledge and is not covered here. It is included here to illustrate the large variety of data needs and analytics needed to identify the causes of the crisis and the different responses. Again, vast quantities of data are involved; they must be analysed and presented for decision.

### 9.7.2 Response to Epidemics and Pandemics

Epidemics have been a common disruption over many years. They range from localized events as for example EBOLA recently in Africa in 2014, or SARS outbreak in 2002, but of more concern is the COVID-19 pandemic. Each of these are followed with careful analysis and recommendations for mitigation to reduce the severity of any subsequent occurrence of the disease.

**The EBOLA Epidemic in 2014**

Ebola has recurred several times, especially in central Africa, with one of the deadliest occurrences in 2014 (see references).

Each such occurrence has been followed with an analysis and recommendations about how to mitigate against future occurrences. These include:

- The importance of quarantining people to prevent spread.
- Distributing information to affected populations of where the spread is occurring and avoiding travel to these areas.
- Developing medical knowledge and response to reduce the severity of the outcome.

One conclusion that is emerging from response to disasters is the importance of capturing and analysing data on how the disease progresses and effective social and medical actions to reduce severity.

Stresses the need of disease mapping and timing to prevent spread (see references). This was illustrated in the COVID-19 pandemic in identifying hotspots through testing and discouraging travel to such hotspots.

Data has been recognized as important in pandemics or epidemics before COVID. The COVID-19 pandemic has provided new challenges because of spread by asymptomatic people, and the rapidity of the spread and the ability of the virus to develop new strains. Different nations respond in different ways to the pandemic. Responses include lockdowns and social distancing, and their effect on transmission is continually evaluated. Extensive data is being analysed to determine the response of ways to manage lockdowns and economic support.

### 9.7.3 Some General Lessons Learned

There are many criteria now emerging as increasingly needed to effectively address disruption. Some are as follows:

- Focus on essentials needed to directly address a disruption.
- Quick response sharing data where necessary.

- Community focus.
- Sharing knowledge with community.
- Unified leadership.
- Questions about mobility to centres of disruption.

## 9.8 Effect on Business

One important consequence of disasters is their effect on business, especially on global supply chains. Global supply chains, as described earlier in ▶ Chap. 8, are organizations where different businesses contribute to a product or service from its inception to the delivery to the ultimate user. A disruption at any point of a supply chain can affect businesses in the supply chain.

Most of the disruptions described in this chapter can lead to disruptions to the supply chain. Floods, earthquakes, and fires can all lead to local supply chain problems. Typical global disruptions especially related in the COVID-19 pandemic can be found in the literature and commercial sites. For example, BigCommerce or IndustryStar among others (see references) include the following:

- Sudden reductions in demand as was the case for the hospitality industry during the COVID-19 pandemic.
- Sudden surges in demand, as again occurring during COVID-19 initially for ventilators, but later for vaccines, and strangely but not surprisingly, young puppies for company during COVID-19 restrictions.
- Reduced productivity through lack of training or unavailability of resources.
- Problems with products requiring frequent attention.
- Price fluctuations as new markets emerge.

There is some consensus that organizations should develop mitigation plans to minimize the effect of disruptions to supply chains. Mitigative actions can include the following:

- Build-up of inventory with consequent impact on storage costs.
- Identify backups.
- Diversity in customers and sources, again leading to costs in maintaining communication links.

Supply chain management itself is a science on its own and readers are referred to the literature for detailed solutions. What is important is to address them in any transformation.

## 9.9 What Happens After a Disruption

Recovery is one of the last stages to deal with disasters. The question now is, do we recover to where we were before the disaster or does a new system emerge? If a building burns down, should it be rebuilt to the same plan, or do we see it as an opportunity to create a new building that satisfies new needs.

It is often due to a disaster that we learn to do things differently. For example, learning how to work remotely during the COVID-19 pandemic.

## Summary

This chapter described disruptions as one of the characteristics of complex systems. This chapter outlined the common processes to deal with disruption distinguishing between mitigation, response, and recovery. It then defined generic processes that address response through focusing on emerging and dominant values that emerge through disruption and the importance of creating expert communities that bring together the expertise that directly addresses the threatened values.

The other important lesson is the importance of data in dealing with major disruptions. It is not only needed to manage responses to reduce damage, but also to analyse the progress of a disruption and experiences in different ways to response. This analysis can then become the lessons learned to both apply in developing mitigation strategies and better ways to respond.

## Exercise

### Exercise 9.1

Go back to the solutions you proposed earlier in the three case studies. Extend them to include resilience to disruption. Use a table like ◘ Table 9.3 to outline how your response to each question would look like. Search to determine risks to natural disasters in your area and how to mitigate to prevent damage. For example:

How would each of the case studies include resilience to a natural disaster like a flood? Or a disruption like a fire?

- What is the mitigation you propose?
- What processes should be followed during the disruption?
- Do you have a recovery plan?
- What additional capabilities would you add to provide a solution that makes your restaurant, shopping center, or supply chain more resilient to disruption?

Use the guidelines in ◘ Fig. 9.1 to identify ways to make your food supply chain, restaurant, or shopping centre resilient to disruption. You can also use the questions in ◘ Table 9.3 to add to the practices you used earlier.

Illustrate your solution with examples. For example, do you think hoarding (which is building up inventories) at any stage of the supply chain results in benefits? Who benefits?

### Exercise 9.2

Find the checklists that describe how to improve safety in the area you live or work in. What would the dangers be and how do you counter them? How would technology help? What data is needed?

Does your office or building have a safety policy? Do you have safety checklist?

What are typical data requirements in mitigation planning for floods? What data would you need to reduce damage during a response?

**Exercise 9.3**

In each case, look at emerging technologies and see how they could help to support disaster management? For example, can AI be used during response to predict the progress of a disruption? And then advise how to mitigate against such progress. Or can you use big data technologies to develop ways to mitigate against disaster in a particular area?

## Reference

Choi, K. Y., & Cheung, R. W. (2013). Landslide disaster prevention and mitigation through works in Hong Kong. *Journal of Rock Mechanics and Geotechnical E, R.W.M. Engineering, 5*(5), 354–365.

## Further Reading

Borland, H., Lindgreen, A., & Maon, I. (2019). *Business strategies for sustainability*. Routledge.
Teece, D., Peteraf, M., & Leih, S. (2016). Dynamic capabilities and organizational agility: Risk, uncertainty, and strategy in the innovation economy. *California Management Review, 58*(4), 13–35.
Wallace, M., & Lebber, L. (2018). *The disaster recovery handbook*. American Management Association.

### References to Practice

Advice on floods in the United Kingdom can be found on www.knowyourfloodrisk.co.uk/flood-advice-guidance
BigCommerce: bigcommerce.com
Ebola importance of data. https://onlinedegrees.unr.edu/blog/the-role-of-big-data-in-global-epidemics/
Ebola outbreak in Africa, 2014: cdc.gov/vhf/ebola/history/2014-2016-outbreak/index.html
FEMA (Federal Emergency Management Service) https://www.fema.gov/pdf/hazard/wildfire/wdfrdam.pdf. Last accessed March 11.
Industry start: https://www.industrystar.com/
Investopedia: www.investopedia.com/news/10-years-later-lessons-financial-crisis/
Rural Fire Service, NSW, www.rfs.nsw.gov.au/resources/bush-fire-survival-plan; www.rfs.nsw.gov.au/__data/assets/pdf_file/0003/4557/Fast-Fact-Evacuation-Plans.pdf. Last accessed March 11, 2021.

# Impact of Climate Change

Contents




I. T. Hawryszkiewycz, *Transforming Organizations in Disruptive Environments*,
https://doi.org/10.1007/978-981-16-1453-8_10

Now we extend solutions to address climate change. This chapter does not address a global solution to climate change, such as what global emission targets should be, but it raises awareness of issues to be addressed by any transformation to mitigate the effects of climate change. This chapter is introductory and readers working in an industry may need to read further to develop more specific and detailed questions. This chapter provides the questions to ask to extend your solutions to mitigate against climate change.

**Learning Objectives**

- Causes of climate change
- Knowledge needs to mitigate effects of climate change
- Including climate change mitigation to transformation
- Extending transformations to minimize contribution to climate change
- How technology can improve response

## 10.1 Introduction

One of the most intractable problems today is that of climate change. The importance to address climate gained prominence in the late 1990s (Gore, 2008) and has developed into a science since that time with the growth of evidence of its impact on the planet (NASA Global Climate Change, 2019). Climate change is increasingly seen as a looming threat to the well-being of people. Its negative effects on businesses and cities have been widely described. As introduced earlier in ▶ Chap. 7, businesses increasingly need to include sustainability as a value, as part of the triple bottom line. In this chapter, climate change is not seen as a problem on its own. It is seen more in terms of the problems that arise from climate change—it thus presents a challenge to any transformation.

10

The challenge is two-fold—how to transform businesses and industries, in ways that any transformation:

- Does not add to the causes of climate change.
- Mitigates the impact of climate change.

One of the most contentious issues today are the causes of climate change. In a book such as this, the emphasis is not to question the science of climate change. It is to reduce the impact of any transformation on climate change and to cater for disruptions caused by climate change. Climate change affects the ability of cities and businesses to raise their values. As a result, businesses are increasingly responding to address the effects of climate change.

Addressing the challenges also provides business opportunities. Increased sensitivity to climate change and its negative impact is now increasingly leading to businesses adopting processes that reduce emissions. It is also creating new businesses that develop products, such as solar panels, that directly replace existing processes and lead to new and better products. The other broad business opportunity is in the management of supply chains that are almost indispensable in the delivery of goods and services from their origin, through production process to ultimate human consumption.

## 10.2 What Do People Value in Their Environment?

The book defined business and city values earlier in ▶ Chaps. 3 and 4. Now we go further and define planet values. These are additional values that are added to stakeholder values. In this section, we add to the values described in earlier chapters and the impact of climate change on these values as shown in ◘ Table 10.1. These can serve to define the best practice questions in design to cover the planet components of the three-bottom line.

Thus, apart from business and city values, there is now greater emphasis on environmental values. Some such values are shown in ◘ Table 10.1.

The next question is the effect of climate change on these values. To describe the effects on values, we go back to the process defined in earlier chapters and begin by capturing stories related to climate change.

◘ **Table 10.1** Values threatened by climate change

| Important environmental values | Some questions to ask |
|---|---|
| Reduced number of natural disasters | Are the numbers of natural disasters increasing? |
| A green environment | Are the number of trees increasing?<br>Are there more green spaces?<br>Are hiking trails available? |
| Reducing waste | Is there a lot of waste?<br>Is waste being put to good use?<br>Is there any unnecessary consumption? |
| Good health support | Is pollution being reduced?<br>Is there access to clean water? |
| Deal with surprises | Are supply chains secure from natural disasters? |
| Biodiversity | Are animal species decreasing? |
| Secure food supplies | Are you happy with food quality?<br>Do you have easy access to food products that you like?<br>Is there degradation of the soil? |
| Secure water supplies | Do you have access to sufficient water?<br>Is the water drinkable? |
| Land availability | Is there land available for agriculture?<br>Is soil quality adequate? |
| Secure housing and shelter | Is your house protected against disruptions caused by climate change? |
| Stable and predictable business conditions | Are business regulations unpredictable?<br>Are supply chains secure? |

## 10.3 What Are the Stories?

### 10.3.1 Natural Disaster of Increasing Frequency

The kinds of natural disasters arising from climate change are widespread in type and scope as illustrated in Table 10.2. Stories are now continually emerging of natural disasters attributed to climate change. It is becoming increasingly apparent that climate change is contributing to their causes.

**Table 10.2** Effects of climate change

| What is happening | Examples |
|---|---|
| Increasing bushfires  | Bushfires in Australia in the 2019–2020 season were the fiercest in many years. So were the US wildfires in 2020 |
| Increasing pollution   | Heavy smoke from fires adding to pollution. The smoke from the fires leads to dangerous pollution in the city. The same was the case in San Francisco in 2020 during the wildfires on the US west coast<br>Heavy traffic is another example |
| Melting glaciers  | Causing Sea level rises. Sea levels have risen around 20 cm. Since 1900 and the rises are now accelerating because of global warming |
| Floods can lead to heavy damage of property, loss of life, and reduction of tourism.  | As an example, floods in Venice, Italy, and Greece with consequent impact on tourism |

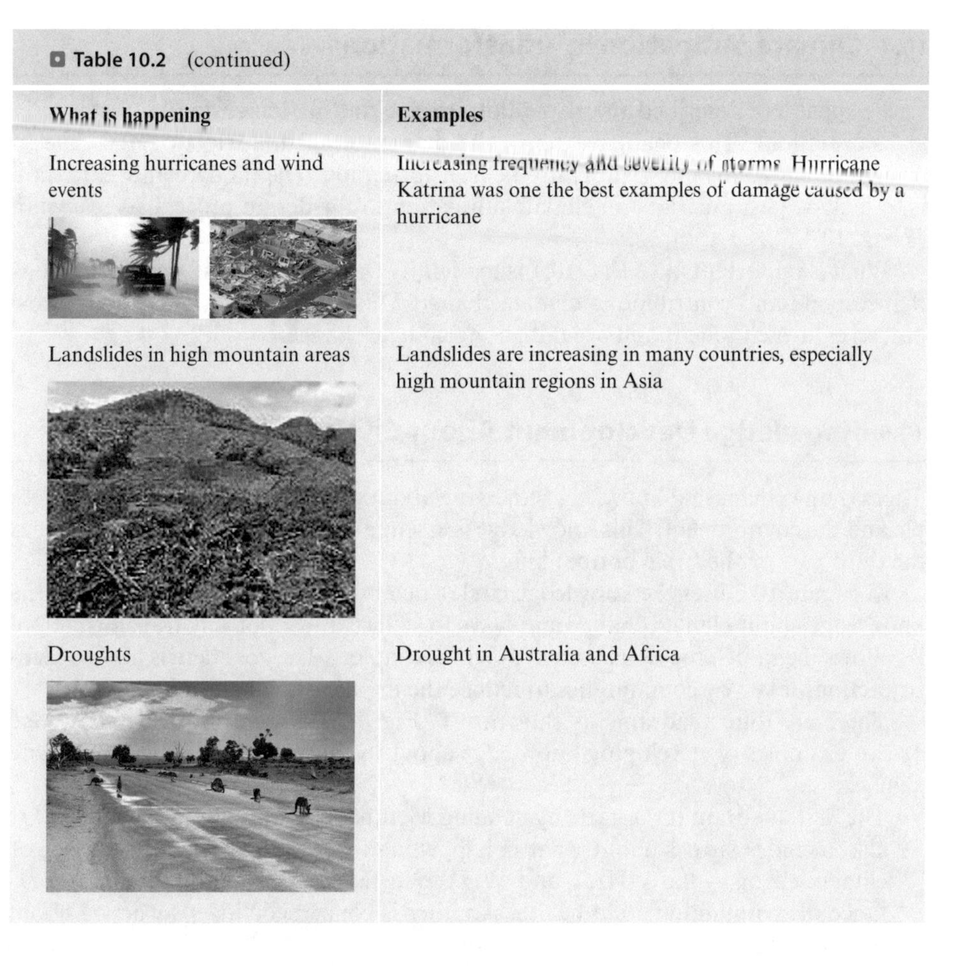

**Table 10.2** (continued)

| What is happening | Examples |
|---|---|
| Increasing hurricanes and wind events | Increasing frequency and severity of storms. Hurricane Katrina was one the best examples of damage caused by a hurricane |
| Landslides in high mountain areas | Landslides are increasing in many countries, especially high mountain regions in Asia |
| Droughts | Drought in Australia and Africa |

One might question whether these disruptions are because of climate change. It is not the intention of this book to question the science, but to make designers aware that these events are happening and to mitigate against them. The argument that climate change is one of the causes of increase is well documented in other literature (Gore, 2008).

## 10.3.2 Efforts of Mitigation

It is obvious from the stories in Table 10.2 that increasingly the environmental values in Table 10.2 are disrupted by natural events. As a result, transformation is increasingly placing emphasis on mitigation to climate change as part of the triple bottom line. Mitigation now includes two goals; ensuring minimal damage, but also mitigation to reduce contribution to climate change and thus reducing the number and severity of disasters.

## 10.4 Climate Mitigation in Transformations

Earlier chapters described the triple bottom line and its increasing importance in transformations. This chapter extends the discussion on the triple bottom line to ways to integrate the mitigation into the transformation. The fundamental extension is to include best practices on climate mitigation to the design process, as shown in ◻ Fig. 10.1.

What is important in ◻ Fig. 10.1 is to identify those parts of the transformations that respond and contribute to climate change. This calls for knowledge about how processes at each touchpoint are mitigating climate change.

## 10.5 Knowledge Development About Climate Change

There is now increasing knowledge emerging about climate change, its effect on people and the environment. This knowledge is essential in addressing climate change as the third part of the triple bottom line.

◻ Figure 10.2 uses the knowledge circle, used in earlier chapters, to organize the knowledge about climate change and ways to reduce impact of any transformation. It follows the same process as used in previous chapters. The goal here is how to identify actions taken by communities to reduce the impact of climate change.

There are four quadrants as shown in ◻ Fig. 10.2. Figure illustrates only one or two examples of developing knowledge about the impact of activities on climate change.

- The first quadrant (Q1) starts by defining what people value about the planet. Q1 also includes stories about what is happening now and the negative impact of climate change—the WHAT and WHY questions. One example is the importance of maintaining good health. Another is concerned citizens worried about the emissions from waste in agriculture.
- The second quadrant (Q2) follows the design process by identifying issues and causes of why values are not met. The example here is the impact of pollution on good health. It also develops ideas of what needs to be done to address the causes—the WHAT IF questions in this case ways to reduce traffic and encouraging the use of electric cars.
- The third quadrant (Q3) implements solutions to the problems identified in Q2. Thus, for example, travel is found to often have a negative impact on climate change because of emission of gases using transportation systems. In most cases, the impact on climate change is not direct—it is the consequences of what people do, rather than the actual actions that lead to disputes. One solution here is to

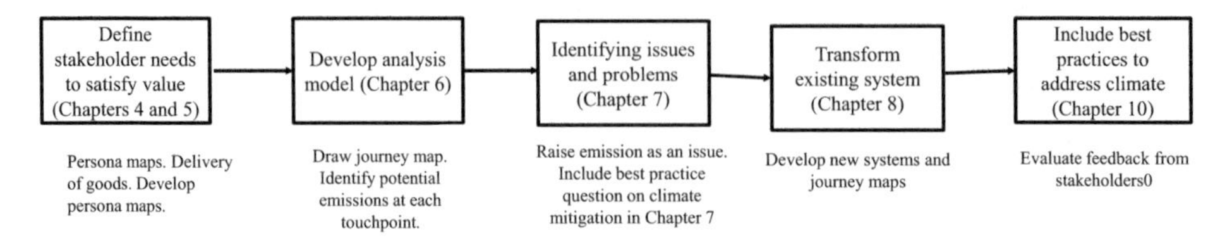


◻ **Fig. 10.1** Including climate change mitigation in transformation

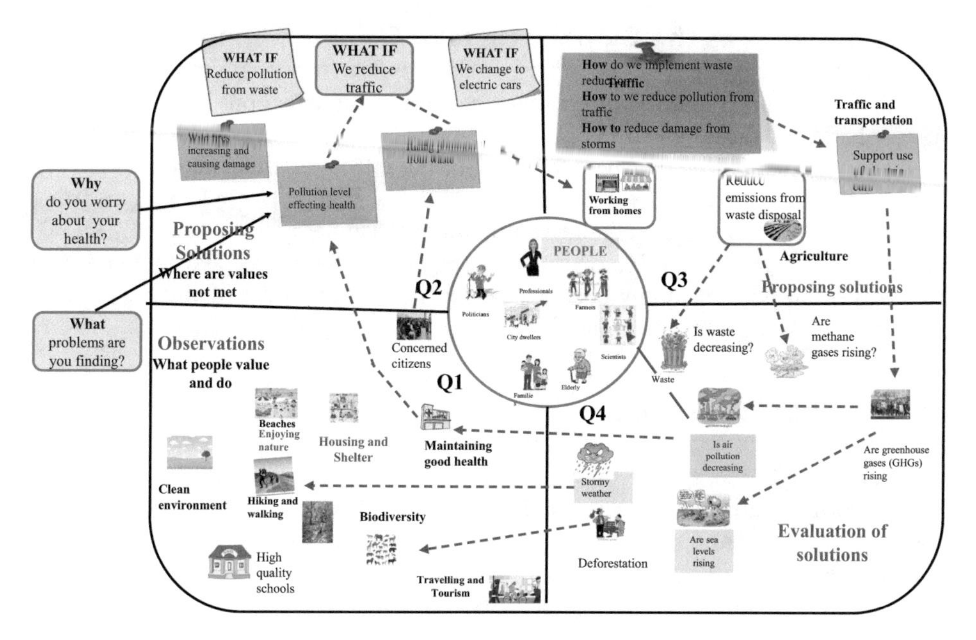


**Fig. 10.2** Knowledge about climate change

reduce traffic by having more people work from home. The other is to support the use of electric cars.

- The fourth quadrant (Q4) looks at the impact of any actions proposed or taken. For example, is the amount of methane gas emissions from agricultural waste rising or falling; are pollution levels rising or falling?

Figure 10.2 shows the typical nature of cycles found in climate change. For example:

Q1—shows that people value good health.

Q2—shows that pollution affects good health.

Q3—looks at best practices to reduce the contribution of traffic, travel to pollution.

Q4—monitors whether the changes made in Q3 reduce pollution.

Figure 10.2 only shows one such cycle. In the study of climate change, there is a multitude of such cycles, thus illustrating the complexity of addressing issues in climate change.

The circular nature of the process is that businesses need to somehow balance creating value for their clients, while minimizing their contribution to emissions and consequently the disruptions caused by climate change.

The focus, however, is not on individual business, but more on industries in which they operate. Typical industries are agriculture, energy generation, transport, supply chains, and so on. Each of these cause emissions that lead to climate change as well as creating value to stakeholders.

## 10.6 Emissions as the Global Causes of Climate Change

Any discussion on climate change cannot get away from emissions of gases into the atmosphere as a cause. The discussion on climate is often clouded by the fact that it is not the direct actions that people take, but the effects of these actions—namely, the generation of GHG (Green House Gases) gases. Whether people, by generating GHG, contribute to climate change is often under dispute by those generally known as climate change deniers. However, increasingly it is being accepted that emission reduction is important in any business design. In the remainder of this chapter, we identify ways that transformation places importance on emission reduction.

### 10.6.1 Global Emissions

It is increasingly recognized that gases, known as greenhouse gases (GHG), have been shown to radiate heat generated on the earth back to the earth, causing the temperature to rise and result in what is commonly called global warming. The main GHG gases are:

- Carbon Dioxide $CO_2$ mainly from energy production
- Methane, mainly from agriculture
- Ozone and Fluorinated Gas (F-Gas)
- Nitrous oxide from soils



The proportion of each of these gases has been extensively studied and one example is shown in Fig. 10.3 from the EPA. The source is: IPCC (2014) *Contribution of Working Group III to the Fifth Assessment Report of the Intergovernmental Panel on Climate Change*.

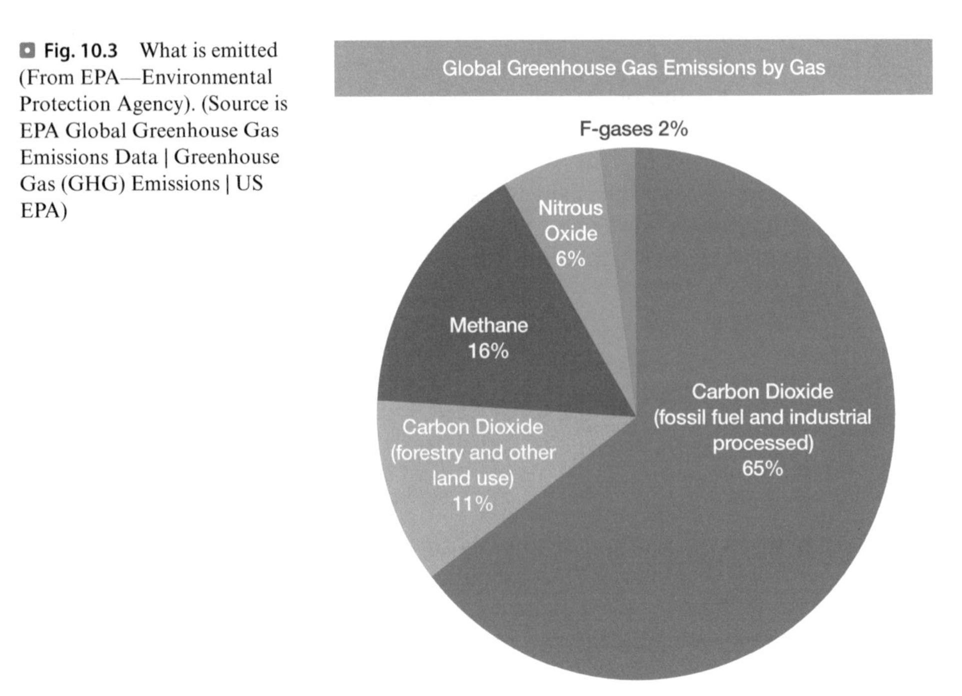


**Fig. 10.3** What is emitted (From EPA—Environmental Protection Agency). (Source is EPA Global Greenhouse Gas Emissions Data | Greenhouse Gas (GHG) Emissions | US EPA)

The goal of any new transformation that supports sustainability must now show that it minimizes the emission of GHG. Different industries often contribute to the emission of different gases. Extensive statistics are available on the kinds of emissions arising from different industries. Again, these have been extensively studied, especially the contribution of different industries to GHG emissions.

### 10.6.2 What Industries Contribute?

Figure 10.4 derived from the Food and Agriculture Organization of the United Nations (FAO) shows the contribution to emissions by industry. Then we ask how to ensure that the transformation adds the minimum to such contribution. For example, buildings should have standards, and use of solar energy whenever possible.

Any such transformation depends on the industry. Following is a brief overview of selected sectors with an overview of the issues raised in the context of climate change. The objective of any transformation is to create systems that are resilient to change and reduce the carbon footprint.

## 10.7 Identifying Causes

Continuing the transformation process. Fishbone diagrams can be used to identify industries that contribute to emissions. Figure 10.5 shows the major contributors to CHG gas emissions (NASA Report, 2019). The use of fishbone diagrams here is

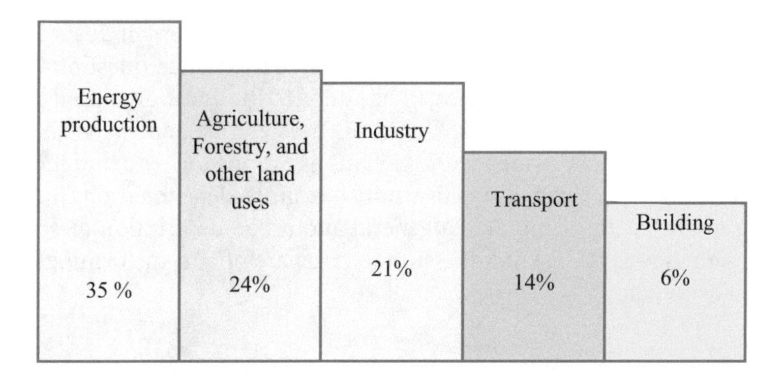


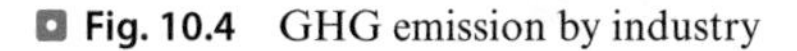
**Fig. 10.4** GHG emission by industry

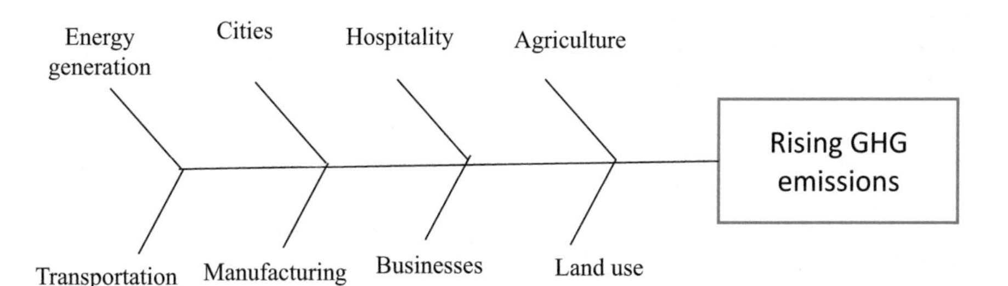


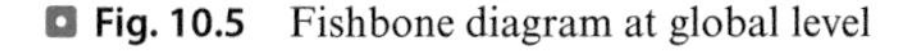
**Fig. 10.5** Fishbone diagram at global level

Table 10.3 General issues

| What needs to be done | Questions to ask |
|---|---|
| Reduce greenhouse emissions from travel | Are you minimizing travel in the transformed business?<br>Are you adopting use of low emission vehicles?<br>Does your transformation support work from home? |
| Reduce CO2 emissions from energy use | Are you using renewable energy whenever possible?<br>Adopting low carbon practices? |
| Maintain a green environment | Is the transformation adding to the green environment?<br>Are you planting trees? |
| Reduce waste | Are you using degradable products whenever possible?<br>Do you contribute to recycling?<br>Are you recovering rare metals from waste?<br>What happens to food waste? |
| Consume what is needed | Are you contributing to a reduction of waste?<br>Does the transformation minimize use of resources whenever possible? |
| Maintain biodiversity | Are you protecting animal, bird, and sea species? |



only exemplary. More detailed fishbone diagrams can be used to indicate the causes at industry levels.

There are many available reports that describe the impact of industries on climate change. Energy and transportation, however, are common in many industries. This book thus sees these two as causes that must be addressed in any industry.

The objective of this chapter is to add some best practice questions to extend design to the triple bottom line. These can be added to the questions asked in ▶ Chap. 7 to identify issues to be addressed. Questions here can be general to the industry, or specific to the transformation. Some general questions for reducing footprint are shown in Table 10.3. There are many more questions dependent on the industry.

The remainder of this chapter is a general and broad description of some industry sectors and indicates the kind of issues to be aware of if transforming businesses in a particular industry.

## 10.8 What to Consider in Transformative Design

It is not possible in one chapter to cover ways to reduce emissions in every industry. The goal is to identify different sectors and issues in those sectors. When transforming a business, it then becomes necessary to:

- Identify the industry sector of the business you are transforming.
- Identify business activities that are adding to the causes of climate change.
- Identify the business issues in that business.

- Do these issues affect or are affected by your business?
- Use the methods described earlier to develop creative and innovative ways to mitigate climate change.

Mitigating climate change is becoming increasingly important and is being addressed in many reports Such reports often focus on industries and you should refer to these when planning transformations.

There are also many reports outlining the kind of mitigations needed to reduce gas emissions. One example is future planet that has several suggestions.

## 10.9 Brief Overviews of Industry Sectors

It is not the purpose of this book to only address climate change, but for any transformation to see it as an important part of the triple bottom line. There are now reports and books on ways to develop strategies or policies for climate change as, for example, (Harvey et al., 2018). What is important for transformations to minimize contribution to emissions by identifying mitigating actions in transformations?

What is needed is that any transformation results in increasing resilience to the effects of climate change, and reduces GHG emissions. The remainder of this chapter briefly describes some sectors and provides guidelines of the major causes of their carbon footprints and potential mitigation strategies. There are two sectors that contribute to most other sectors—transportation and energy use, and must often be considered in any transformation that follows the triple bottom line.

### 10.9.1 Energy Use and Generation

Energy use is a part of any business or organization. The goal is to minimize energy use and thus reduce emissions. Energy management is an important consideration in transformations that address the triple bottom line. Issues here are ways to ensure use of energy, look at ways to organize processes, and restructure to minimize energy needs. The questions asked here in any transformation are as follows:

- Does your transformation minimize energy use?
- Are you using energy produced with minimal GHG emissions?
- Are you increasingly using energy from renewable sources?
- Does your transformation include an energy management activity?

Most people are aware that the way energy is generated is causing considerable discussion—especially in the use of coal to generate energy.

### 10.9.2 Transportation

Like energy, transportation is essential to modern life. People need to travel to work, shopping, and entertainment. Travel management, just like energy management, is important in any transformation in any industry. Some examples of mitigation are shown in ◘ Table 10.4.

The questions here include the following:

- In developing journey maps, do you reduce the need to travel?
- Are you reducing gas emissions in your vehicles?
- Do you only travel for essential tasks such as training where hands-on experience is needed? Or is it just for document production?
- Does the transformation include a travel management plan?

### 10.9.3 Business Administrative Processes

A major part of any business is what is typically known as administrative work—keeping accounts, making reports, and distributing information. ◘ Table 10.4, together with fishbone diagrams, can be used here to identify where design should include ways to mitigate the impact of climate change.



Working from home, especially in times of major disruptions such as the COVID-19 pandemic, has led to two mitigation strategies—working from home and reduction in business travel. The response to sudden lockdowns and working from home has been quick and people have quickly adapted to using existing technologies as exemplified in ◘ Table 10.5. General reports are that many tasks can be carried out remotely and productivity has not been severely affected and in some cases remained the same. However, technology innovations no doubt will result in new services that will make work from home more productive. More details of what is needed are described in ▶ Chap. 12.

◘ **Table 10.4** Some causes of emissions in transportation

| Activity leading to emissions | Mitigation |
|---|---|
| Travelling to meetings uses energy that contributes to emissions | Hold meeting remotely whenever possible |
| Unnecessary lighting adds to energy use | Reducing energy use through reducing office space |
| Excessive use of papers ultimately requires more trees to be cut | Working from home to reduce travel to office<br>Increase remote work |
| Excessive elevator use to access services | Reduce elevator use by locating frequently accessed services throughout the building |

**Table 10.5** The many ways to work from home

| Type of work | Mitigation |
|---|---|
| Preparing reports requiring reference to data | Office technology for developing and transmitting files<br>Microsoft Office, for example |
| Holding meetings | Ability to collaborate interactively<br>Using technology such as ZOOM |
| Making presentations | Ability to engage listeners interactively in webinars |
| Taking classes | Presentive interactive classes—ZOOM is one of the most common technologies |

**Table 10.6** Some causes in emissions in hospitality

| Activity leading to emissions | Mitigation |
|---|---|
| Travel to destinations and excessive travel to entertainment | Find ways to reduce journeys while maintaining experiences<br>Develop new local activities<br>Develop travel itineraries that minimize travel |
| Hotel stays result in energy use | Use accommodation with minimal energy use<br>Use renewable energy in hotels<br>Are sensors included to monitor energy use? |
| Large events | Focus more on local sporting events<br>Community entertainment. |

### 10.9.4 Hospitality and Tourism

Hospitality covers many activities—restaurants, accommodation, and travel. Tourism's biggest contribution to climate change is travel, followed by accommodation. Table 10.6 includes some activities that can contribute to emissions and ways to mitigate them.

At the same time, greater focus on family and local communities during the COVID-19 has resulted in people looking closer to their community for entertainment and social events.

### 10.9.5 Cities and Buildings

Cities are significant contributors to emissions in many ways. Traffic and transport are one; building construction and consumption is another. Connectivity in cities has been shown to result in minimizing emissions. Better connectivity basically reduces the amount of travel as people need to go shorter distances to do what they need to do.

**Table 10.7** Examples of causes in emissions from city buildings

| Activity leading to emissions | Mitigation |
|---|---|
| Moving between different buildings can lead to travel with resulting emissions | Provide and use public transport<br>Use technology to hold meeting and move documents |
| Lighting city streets requires energy use | Using solar power whenever possible |

Emissions here include the construction of building and then the emissions of building occupants. It is seen as a growing contributor in the future, especially with the movement of people into cities and the growth of dwelling construction. Since the Paris Agreement, there has been rising activity in limiting emissions from buildings both through technologies used in household utilities and in the design of the building itself (Table 10.7).

### 10.9.6 Agriculture and Food Production



Agriculture is commonly grouped into what are known as AFOLU activities (Agriculture, Forestry, and other Land Use). It includes the following:

- Farms that include crops, animal, and fisheries
- Transport and distribution of products especially maintaining quality in transit
- Land management
- Marketing
- Handling, storage, and processing of food products

Most people find it surprising that food production and distribution is one of the major contributors to greenhouse gas emission. Contributions to GHG emissions in agriculture has been studied for many years. It may be surprising to some that food production contributes to 26% of GHG emissions. Hannah Ritchie, in an article in Our World in Data (see references), describes the contribution to such emissions from food production. A summary of the report is shown in Fig. 10.6, which is derived from her freely available site and illustrates how food production contributes to emissions. Other studies have shown contribution to emissions is around 13% of CO2, 44% of methane, and 82% of nitrous oxide.

As an example, one can look at meat production, which, from Fig. 10.6, contributes 31% of emission caused by food production. Meat production is a complex process (Bernabucci, 2019) that goes through several steps. Each of these steps impacts climate change (Rust, 2019). Most people are not aware that the food production sector contributes 31% of methane emissions worldwide. The contribution happens in several stages of the meat production cycle shown in Table 10.8. This

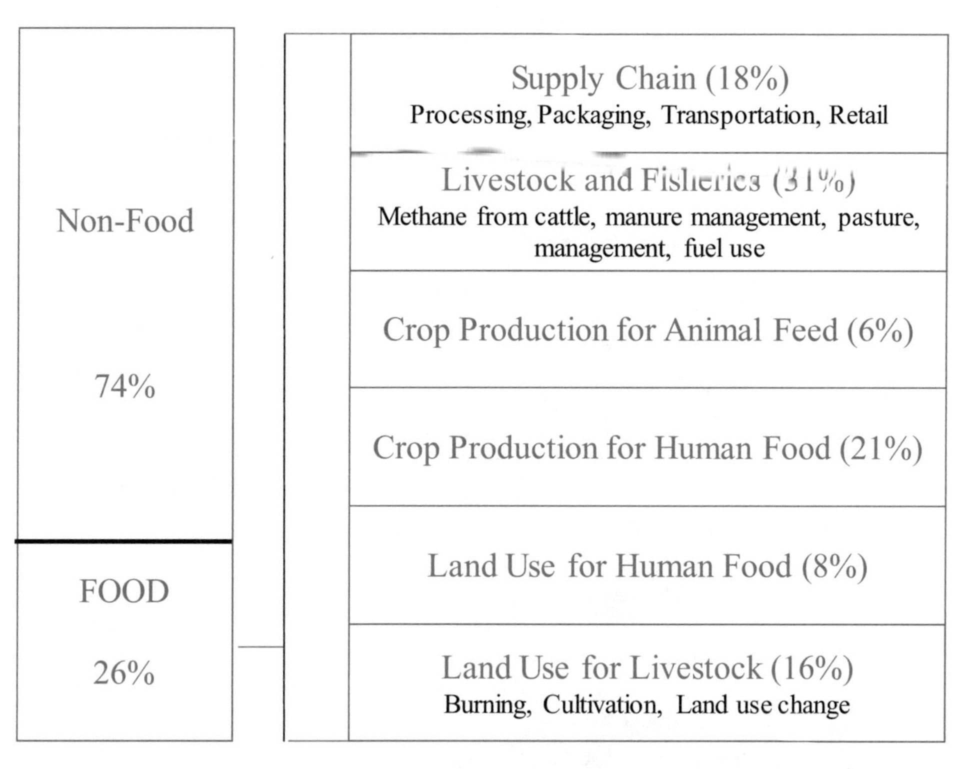


**Fig. 10.6** Contribution of food production to GHG emissions

table serves as a guideline for designers in the meat supply industry to indicate where attention to the triple bottom line is needed.

What, then, are the questions to ask in any transformation in the industry?

- Have you ensured that waste disposal minimizes emissions?
- Have you reduced waste disposal?
- Are you minimizing travel in distribution of food?

Similarly, milk production in dairy farms increasingly contributes to emissions, especially methane gas, similarly to meat production. Dairy cows, similarly to cows bred for meat, have special feeding needs. Some details can be found in the references.

### 10.9.7 Land Use

Land is becoming a scarce resource in an ever-growing population. Land use is a related issue to food production. It is here that issues of land use get increased atten-

**Table 10.8** Meat production

| Activity leading to emissions | Mitigation |
|---|---|
| Feeding animals requires the creation of feedlots, results in generation of methane gas  | Matching animals and feedstock to minimize generation of GHG |
| Processing of animal products. Waste from slaughter and packaging processes. Waste disposal from animals contributes to emissions while wasting  | Is waste properly disposed?<br>Are there provisions for capturing gases as waste decomposes? |
| Distribution through stores and supermarkets and resulting waste during delivery and sale  | Is there careful planning to minimize overstocking and resultant waste on disposal? |
| Preparation of meals and waste disposal  | Is there planning in meals and venues to reduce waste?<br>Is waste disposed properly? |



tion. It comes from two directions—the use of land in agricultural production and land degradation. Thus, any business design should look at both. Issues here include the following:

- Deforestation and loss of trees that reduce absorption of greenhouse gases.
- Land clearing for residential, commercial, and recreational use.
- Land degradation through overuse and climate variation.

Designers in any transformation that includes land should refer to many reports on land degradation and ensure their designs mitigate land degradation. Examples of reports include the following:

- Land degradation in Africa
- Reports stress that land care (Kapur et al., 2011) is important as land in needed to generate food for an ever-growing number of people.
- Using land care itself as mitigation of reducing release of GHG (see references) (Table 10.9).

## 10.9.8 Water Management

Water is now becoming a scarce resource with growing population faced with climate change. Transformation should again look at uses of water management and mitigate to conserve water. Issues here include the following:

- Water being wasted through lack of capture infrastructure.
- Water distribution policies allocating scarce supplies to high use industries, as for example, mining.
- Water being wasted through lack of care in domestic and industry setting.
- Maintaining groundwater quality and supply in dry countries (Table 10.10).

**Table 10.9** Issues in land use

| Activity leading to emissions | Mitigation |
|---|---|
| Land clearing leading to increasing deforestation and desertification | Are more trees planted following land clearing?<br>Are alternate methods like multi-level farms or greenhouses (Fig. 10.7) used to increase land availability? |
| Land overuse | Is sufficient care taken to avoid land degradation? |

**Fig. 10.7** Alternatives to land use

■ **Table 10.10** Major causes of water waste

| Activity leading to water waste | Mitigation |
| --- | --- |
| Water not being captured in rainy seasons | Are sufficient water capture methods used?<br>Are rainwater tanks in use? |
| Allocation of water to high use activities such as mining. | Are there policies in place to distribute water?<br>Is the use of water being monitored?<br>Is its quality being maintained? |
| Excessive use of water | Are there guidelines in use to minimize water usage? |
| Water going to waste in domestic or industry setting | Are there methods in place to ensure water is not wasted?<br>Are there ways to encourage behaviour that reduces water waste? |
| Depleting groundwater requiring increased pumping costs. | Are the best technologies used to pump water?<br>Is groundwater being overused? |
| Maintaining ground water quality. | Are chemicals being dumped in local waste?<br>Is there fertilizer use close to ground water deposits? |



**Summary**

This chapter described the more negative side of impacts of climate change. Its focus is to raise awareness so that designers reduce emissions both in the creation of new artefacts and their subsequent use by consumers. It also showed how to include the effects of climate change in the transformation process. It described how to identify potential emissions in business processes and the type of mitigative actions that can be taken.

It stressed the importance of ensuring that any transformation minimize emissions and mitigate the effects of climate change. It described how industry sectors contribute to emissions and questions that designers should ask to ensure transformation follow best practice in addressing climate change. You might also note that waste management is common in many activities, and waste management is becoming increasingly important in dealing with climate change.

## Exercises

### Question 1

For each of the case studies introduced in ▶ Chap. 1:

- Identify the major industry for the case and develop a fishbone diagram showing causes of emissions, using tables like ■ Table 10.10.
- What kind of mitigation would you propose to reduce emissions?
- You should also try to develop the kind of knowledge cycle using ■ Fig. 10.2.

**Question 2**

You can also elect an industry sector and describe what you feel is the negative effect of climate change and what the industry can do to mitigate these effects. Again, draw a knowledge cycle showing potential for emission reduction.

## References

Literature on climate change is very wide and covers a multitude of issues. Readings in transformations in this area are usually science or industry reports.

Gore, A. (2008). *An inconvenient truth*. Rodale Publishing.

Harvey, H., Orvis, R., & Rissman, J. (2018). *Designing climate solutions: A policy guide to low-carbon energy*. Island Press.

Kapur, S., Eswara, H., & Blum, W. E. H. (2011). Sustainable land management. In *Learning from the past for the future*. Springer.

Rust, J. M. (2019). The impact of climate change on extensive and intensive livestock production systems. *Animal Frontiers, 9*(1), 20–25.

## Further Reading

Dairy farming.: www.bbc.com/future/article/20201208-climate-change-can-dairy-farming-become-sustainable

Food and Agriculture Organization of the United Nations. (2015). Greenhouse Gas Emissions: from Agriculture, Forestry and other Land Use.

Gariano, S. L., & Guzzetti, F. (2016). Landslides in a changing climate. *Earth-Science Reviews, 162*, 227–252.

International Energy Agency. (2017). Towards a zero-emission, efficient, and resilient buildings and construction sector" Global Status Report.

Isla, A. (Ed.). (2019). *Climate chaos: Ecofeminism and the land question*. Janna Publications.

Land degradation in Africa.: http://www.fao.org/docs/eims/upload/288683/LandClimate_Executive_Summary_English.pdf

Mason, S. G., & Fragkias, M. (2018). Metropolitan planning organizations and climate change action. *Urban Climate, 25*, 37–50.

Mead, L. (2019). Sustainable land management critical to combating climate change: IPCC (Intergovernmental Panel on Climate Change) Special Report.

Nettle, R., Ayre, M., Bellin, R., Waller, S., Turner, L., Hall, A., Irvine, L., & Taylor, G. (2015). *Empowering farmers for increased resilience in uncertain times* (pp. 843–855). Animal Production Science.

Redante, R. C., Medeiros, J., Vidor, G., Cruz, C. M. L., & Ribero, J. L. D. (2019). Sustainable production and green product development: Using design thinking: Promote stakeholders' engagement. *Sustainable Production and Consumption, 19*, 247–225.

Reinhardt, F. L., & Toffel, M. W. (2017). Managing climate change: Lessons from the U.S. Navy. *Harvard Business Review, July–August*, 103–112.

Ritchie, H. (2019). Food production is responsible for one-quarter of the world's greenhouse gas emissions, November 06, 2019. ourworldindata.org/food-ghg-emissions.

UN Environment and International Energy Agency. (2017). Towards a zero-emission, efficient, and resilient building, and Construction Sector. Global Status Report 2017.6.

Wellenger, P. (2018). *Smart buildings: Four considerations for creating people-centered smart, digital workplaces*. Deloitte Insights.

## Web References and Industry Reports

Bernabucci, U. (2019). Climate change: impact on livestock and how can we adapt. *Animal Frontiers, 9*(1), 3–5. https://doi.org/10.1093/af/vfy039

EPA (2014). Global Greenhouse Gas Emission Data. www.epa.gov/ghemissions/global-gas-emossion-data.

NASA Global Climate Change: Vital Signs for the Planet; The causes (2019). https://climate.nasa.gov/causes. Last accessed 20 Nov. 2020.

# Making Smart Cities Liveable

Contents




I. T. Hawryszkiewycz, *Transforming Organizations in Disruptive Environments*,
https://doi.org/10.1007/978-981-16-1453-8_11

Many businesses operate within a city or within communities. Cities both develop services to improve liveability in the city, while at the same time, cities develop services that help businesses in the city to create business value. City design must develop the infrastructure that enables businesses to operate in ways that meet business goals, while at the same time, making it easier to meet city values. To do this, cities develop policies to enable businesses to add to their business value and to provide services to support these policies.

**Learning Objectives**

- What are smart cities?
- Cities as organizations
- Policies to create smart cities
- Ways to create policies
- Services to implement policies

## 11.1 Introduction

More and more people live in cities. Design of cities follows the same ideas as in earlier chapters by identifying needs and issues of why they are not met. City development itself must consider the needs of business and its citizens, while at the same time mitigating the effects of disruptions, and increasingly the effects of climate change.

City jurisdictions, however, do not want to directly run businesses and social activities. What they want to do is to facilitate ways for others to do so. Any transformation should consider its impact on city values (defined in ► Table 4.6) and identify what the city should provide to help realize personal and business values.



- Developing the infrastructure that enables businesses and entertainment to flourish.
- Developing community spaces for people to participate in creative activities.
- Being able to deal with disruptions.
- Ensuring sustainability and reducing impact on climate change.
- Providing the services needed by citizens.

This chapter begins by describing the ideas behind smart cities prior the COVID-19 disruption and follows how the characteristics that define a smart city can be impacted by the pandemic. In summary, the generally accepted city goal as seen in this chapter is to "develop an environment for businesses to thrive and citizens to enjoy their life, while satisfying the triple bottom line."

## 11.2 What Are Smart Cities

The definition of a smart city (Song et.al,) is still emerging. In summary, there are three major strategic approaches:

- Biswas (2019) emphasizes *inclusiveness of city citizens*, Biswas analysed two cities in India (Chennai and Kolkata) using this framework. Inclusiveness here includes the ability of citizens to participate in decision making (governance, physical planning), be provided with opportunities (to buy land, obtain shelter), and share the benefits (community infrastructure, productive employment). Rebernik (2019) also stresses inclusiveness and the need to empathize with citizens to identify and provide needed services.
- Another definition is that a smart city is an *urban ecosystem* that places emphasis on the use of digital technology, shared knowledge, and cohesive processes to underpin citizen benefits in themes such as mobility, public safety, health, and productivity. The term assets is often here to describe what jurisdictions provide for communities, for example, hospitals, roads, ports, are city assets. There is now greater emphasis on technologies such as telecommunications and information systems as assets. These can include support of Wi-Fi, open data, or monitoring activities to support safety as assets that can support communities. Increasingly questions are being asked whether assets are to be provided by private or public sectors, or both working together.
- Still another definition is that a city is a *collection of smart connected communities*. Large cities (sometimes called megacities) that are becoming smart but also regional towns and districts also are introducing smart services for their citizens. During the last few years, almost every city has some smart services that it provides for its citizens. These also include regional cities; one example is Port Augusta in South Australia, which is becoming increasingly totally reliant on solar energy.

A smart city is also often seen as a combination of these three directions. Recently, for example, Appio et al. (2019) identified a definition of a smart city as one that encourages "competitiveness of local communities through innovation while increasing quality of life for its citizens through better public services and a cleaner environment."

A smart city can be perceived in several ways.

- Communities continually innovating to improve their quality of life within the infrastructure provided to them.
- A partnership between city authorities, who provide the assets, and communities, who use the assets to improve their well-being.
- Creating assets that can be used by communities to improve their well-being in innovative ways.

Creation of assets or alternatively infrastructure can be defined as roads as was the case to support the suburban sprawl, but increasingly include easy access to Wi-Fi as was one asset to contribute to London's standing as a smart city, and increasingly as open data for use by communities. Thus, rather than seeing the city as a set of independent communities, one goal is to see them as a connected ecosystem (Jacobides et al., 2018).

Zygiaris (2013) defines a few layers through which an asset infrastructure can be developed. It starts with the green level, followed by layers through which assets are developed until an innovation layer, where creativity results in the emergence of new business models that lead to improved quality of life. The intermediate layers include the interconnection layer. Instrumentation layer followed by open integration and shared applications. It is also suggested that such ecosystems can include outlying rural communities (Fennel et al., 2018).

### 11.2.1 How Smart?

Earlier, ► Table 4.2 in ► Chap. 4 defined values to be met by cities and what are known as smart cities. These included measures such as connectivity, safety, and mobility among others. Safety has taken a new meaning in 2020 with the COVID-19 pandemic, where safety is extended to reducing the chances of catching virus from others. Cities are often judged by their rankings, which are independently developed and published. Rankings are easily accessible over the Internet. There are rankings for safe cities, most liveable cities among others. City development is now seen as a worldwide issue and there is extensive discussion on how to develop city policies that result in achieving the city goal. These measures can be used to identify policies to address major uses applicable to a city.

## 11.3 What Are Cities Doing to Become Smart



There are now many cities adopting innovations to become smart. Sometimes cities define their needs in terms of themes, and some themes may be important in some cities but not others. These are sometimes seen as themes that are important to its citizens. One challenge in designing smart services is to identify what theme to address. Many cities are now setting goals for service delivery to citizens. Each city needs a different set of services. Many now have defined strategies that are published on their websites.

**► Examples of Important (Sometimes Dominant) Strategies in Some Cities**

Cities are different as are themes chosen to make them smart. A good summary is found on the Inter-America Development Bank.

There are major differences in smart city development, often dependent on the stage of city development. Perhaps the biggest difference is between older cities in Europe and the US, and the emerging development in Asia. The main difference is that, in many existing cities, smart communities are developed through change made to existing infrastructure to create the built environment conducive to innovative communities. Many large megacities see an alternate where, rather than changing internal infrastructure, outlying cities are built, with fast connections to the megacity centre. Communities established in the outlying centres are chosen to focus on innovation.

**Developments in Existing Cities**

Many existing cities are now developing specialized communities in their existing city, supporting them with complimentary assets. One common is support start-ups, as in Barcelona and Amsterdam. The focus is strongly on supporting communities with becoming a functional city that delivers services to communities. One goal is to prevent social exclusion and increasing participation in social activities. Apart from assets that support health and well-being, including a focus on exercising.

Helsinki, which has the goal of becoming the most functional city in the world, endeavours whenever possible to develop assets for learning. Functional support for providing Choice, Growth, Service (*Health, learning, Responsible management*), Catering for the aging population is a major theme.

**Developments in China**

Beijing,—reducing pollution is an important theme in China. The trend in China often is to develop smart new cities from the ground up. One example is Xiongan close to Beijing. It is proposed as a model city, with no high-rise buildings, and limited to 5 million people, with high-speed travel to major centres, in particular Beijing, with centres with support for innovation. Such centres initially will include branches of leading Universities and companies such as Alibaba and Baidu. Another similar development in China is Yinchuan.

**Some Developments**

Singapore (Lee et al., 2014) focusing on mobility, health. Safety, Productivity.

Masdar City—the major theme here is energy for sustainability. ◄

Many cities now define their plans to become smarter. You can look up what is important to city or its plans for becoming smarter using the World Wide Web.

## 11.4 How Are Cities Transforming

Transformational design of cities is a wicked problem. A transformation should follow the principles and processes defined in earlier chapters. There is no standard solution for a city—each city is different. The needs of each city are also different and so are solutions. From ► Chap. 2, solving wicked problems is untangling a set of strings. These is keeping cities safe, supporting mobility, and a healthy environment.

When transforming cities, the approach is to manage the complexity in two steps, as shown in ◘ Fig. 11.1. There is a two-phase process found in smart city development.

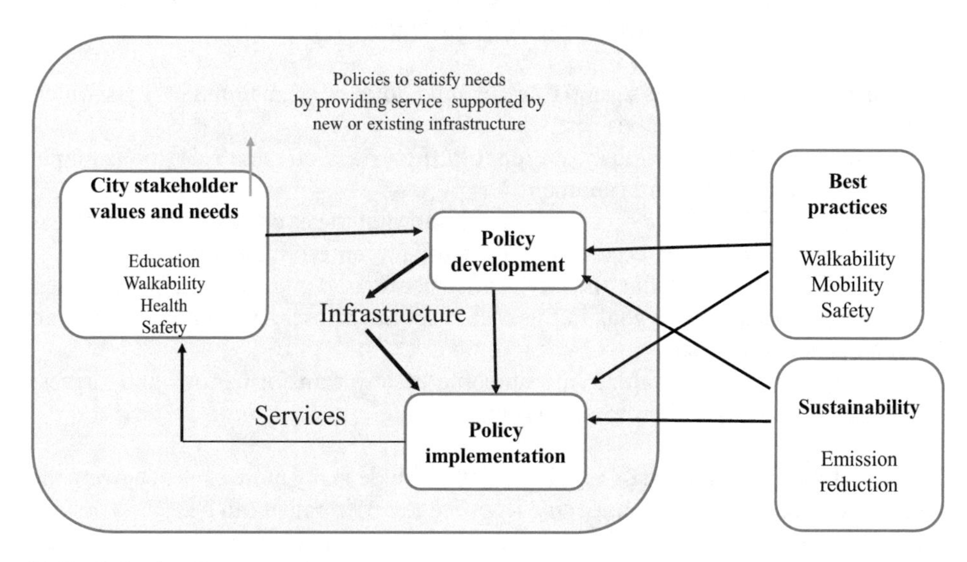


◘ **Fig. 11.1** A policy approach

- First, develop a ranking of major needs to be supported, in fact a strategic policy document.
- Develop and implement the policy to deliver the services while adopting best practices.

For both the policy development and implementation, use the methods described in previous chapters.

Policy development is generally broad and focuses more on policies that address the needs of a large number of people. These may be policies that address citizen values such as a safety policy or a health policy. The infrastructure can, for example, be roads needed to encourage trade, or hospitals to provide health benefits. We can of course see cities as providing services to its dwellers, the same as we see businesses as providing services to its customers. Hence the same design process can be applicable, but in this case, at a city level. Policy developments call for extensive knowledge about a city.

## 11.5 Knowledge for City Transformation

▶ Chapter 4 defined the city values that are increasingly seen as those that are important in raising the quality of life of the growing number of citizens. These include safety, walkability, among others and provide services to all city dwellers rather than individual businesses. We have a checklist in ▶ Table 4.2. This checklist includes general issues like health delivery, safety, among others. These call for services that improve health, provide education, or environments for exercise and entertainment. However, there are many other issues, including economy, such as those identified in a recent McKinsey report on "Smart Cities: Digital for solutions for a more liveable future." These include cost of living, jobs, accommodation, as well as a clean environment. These then become issues that must be addressed by city planners. Solutions often now include ways to use technology and provide it with the necessary data needed to provide services.



Just as a business, the development of liveable cities can be seen in the context of knowledge development. We develop knowledge of what city residents need, what solutions are needed, and how to provide them. One example is shown on our knowledge development circle in ◘ Fig. 11.2.

- The inner circle contains several roles, or stakeholders, often found in cities, which is much larger than in a business.
- In quadrant Q1 as previously, we identify the values of these roles, an example here is schooling and entertainment.
- In quadrant Q2, we develop knowledge about what issues citizens faced in achieving their values, lack of access to schools and few entertainment areas.
- In quadrant Q3, we define the city infrastructure and change infrastructure can be changed to achieve a goal, in this case, a special bus service and designating an entertainment district.
- In quadrant Q4, we evaluate the outcome of any transformations and suggest further changes based on what we learn.

Similarly, to ▶ Fig. 10.2 and ◘ Fig. 11.2 would include many more circles showing the increasing complexity of running a city. Knowledge development can be seen as providing a systematic way to develop services that cities can provide to their citizens. What most cities do is to identify the services needed by dwellers and visitors, set priorities as

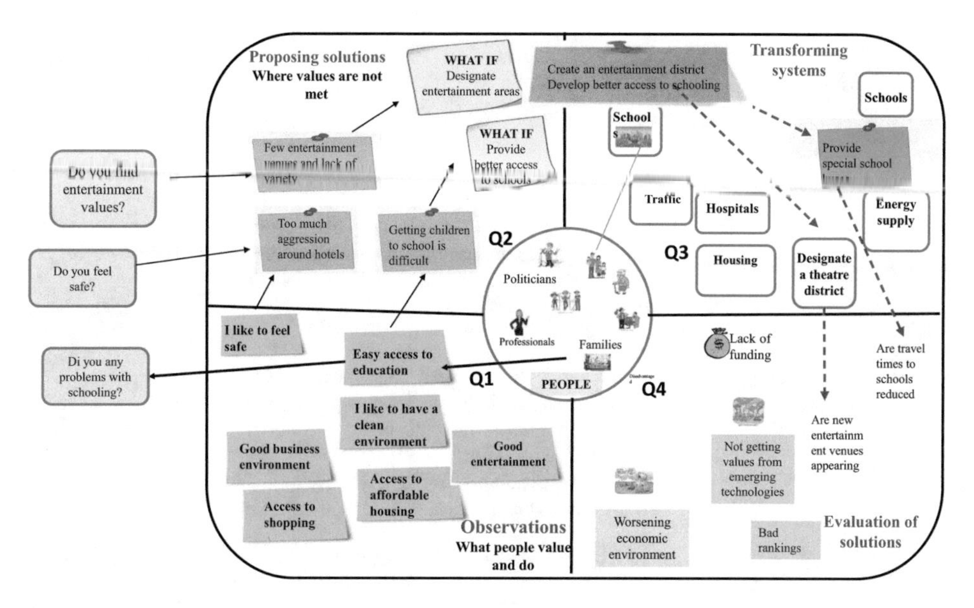


Fig. 11.2 City knowledge

to develop a policy to develop the necessary infrastructure to provide services. These often call for city transformations that can use the tools described in earlier chapters.

The usual way to manage in complex environments, such as a city, is to develop policies. These policies define the services needed by citizens and the infrastructure needed to provide such services.

## 11.6 The Policy Development Process

What are the policies? Policies in general address values defined earlier in ▶ Table 4.6. The question is what policies are needed for your city and how to develop them. Many cities are now adopting policies that improve these values.

### 11.6.1 What Policies?

There are no standard set of policies that are applied to each city. Each addresses its citizens' values. The OECD, for example, makes recommendations (see references). These include recommendations such as:

- Set goals by encouraging citywide planning.
- Encourage dense development.
- Regenerate existing area.
- Reduce congestion.
- Create a greener environment.
- Improve quality of life such as a good walking environment.

Table 11.1 Examples of policies

| City | Example |
|---|---|
| Singapore<br>Transforming through technology | Enabling a culture of innovation and experimentation through making data collected by agencies available, developing an industry start-up ecosystem, and supporting cross-border collaboration. Outlined in more detail in the references which describes transforming Singapore to see key milestones such as improving mobility in more detail while identifying needs of the variety of stakeholders found in any city |
| Helsinki, Finland<br>Most functional city in the world. | The goal here in to become a smart, sustainable city by the sea (see references). This includes the following:<br>Improving mobility while reducing emissions<br>Using clean energy sources<br>Developing a clean and circular economy by reusing waste<br>Making data available through information technology |
| San Francisco USA<br>Developing accessible services to all stakeholders, including residents, visitors, and businesses | Improving service experiences through services, including data sharing, disaster preparedness, and experience through digital services<br>Again, in the references, you can find various strategies and policies to ensure technology adds to city values |



Table 11.1 introduces examples of policies developed in a few cities. These are only a few examples and readers interested in smart city policies are encouraged to get such information, which is widely available on the Internet.

These are only a sample of policies showing differences between cities.

### 11.6.2 Some Policy Examples

Table 11.1 shows some examples.

These guidelines will now no doubt change during the post-COVID-19 development. There is, for example, in Australia, at least a movement of people out of the city to nearby country areas where properties are less expensive and bigger, encouraging family life while working from home. There may be such a movement to smaller communities that can address the challenges of COVID-19 towards behaviours that reduce transmission of the virus.

One of the emerging goals is making city data available to city stakeholders.

### 11.6.3 The Importance of Data

Data is increasingly seen as important in creating smart cities and contributing to smart city measures. The emphasis is on data that is not specific to any one city application but shared throughout the city. For example, collection of information on city movement can benefit traffic flow, police, and health authorities.

The question then to ask is whether city data is being collected for specific applications or is there a city-centric effort to collect and organize data for use throughout the city sometimes known as an open data policy.

The general trend here is that data is being collected but more on an application-centred rather than citywide approach. Furthermore, such data should be open to any business application. Such data can be collections from citizen comments or from city authorities. Data analytics are then seen as essential to analyse any collected data.

### 11.6.4 Identifying Services

City administrations collect large volumes of data on city activities. Data is collected through sensors such as traffic movements or administrative procedures. The question then becomes who can have access to such data, or information derived from it. Both people and businesses can use such data in their planning or everyday activities. For example, knowledge of population trends is useful on where to locate your business and services or products for your customers. It is providing information relevant to individuals. One example of such open data is the air quality dashboard of Bristol (see reference). As shown on ▣ Fig. 11.3, residents in Bristol can see the pollution levels at their selected location.

Most people are now aware of the importance of data in the COVID-19 pandemic. Data on the spread of the Coronavirus is now provided daily and processed to indicate hotspots, or initiate lockdowns.

## 11.7 Policy Development

One way to describe city planning is to divide between policy and implementation. This chapter divides city planning into two parts:

- Identify services needed by city stakeholders.
- Developing the infrastructure to deliver the services.

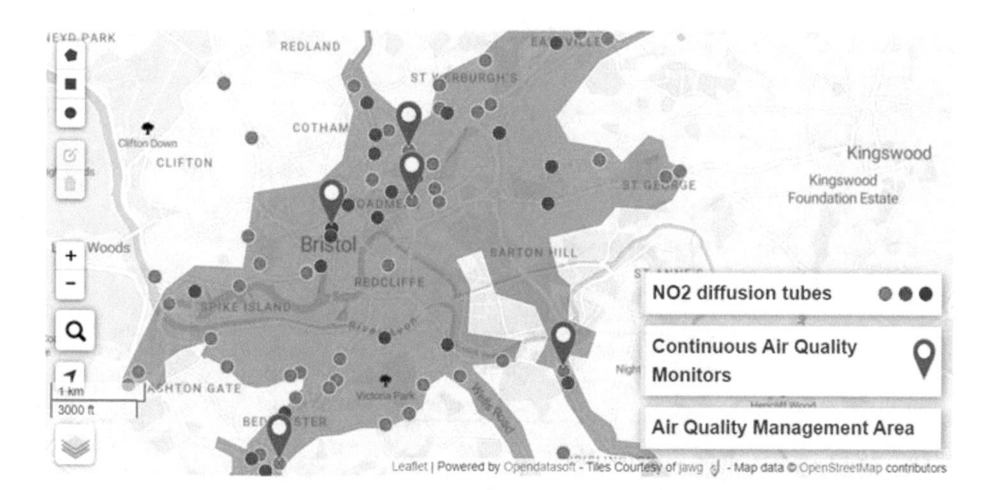


▣ **Fig. 11.3** Showing air quality

### 11.7.1 How to Decide and Prioritize?

Policy development often follows the same process as in earlier chapters, although at a higher level. As shown in ◘ Fig. 11.4, the first activity is to identify stakeholder (city dweller and visitor) needs. Most cities have a wide range of needs. Consequently, one of the first steps is to identify them and set priorities. Then develop a policy to deliver the physical and administrative infrastructure to deliver services to satisfy these needs. The activities here, as shown in ◘ Fig. 11.4, are to perform the following:

- Identify major community needs.
- Set policies to develop services to meet these needs.
- Develop the supporting infrastructure.
- Provide services to support the policy.

### 11.7.2 Identify Stakeholder Needs and Values

The transformation process described earlier for businesses can also be used here. Thus, the design now focuses more on city values rather than business values. At the same time, stakeholder values continue to play an important role. We are now dealing with the needs of wider communities and ways to provide services that support their needs. These depend on the type of community and stage of its development.

Fundamentally, the process follows a value-based approach described in ▶ Chap. 8—we first find the knowledge of what people need, what are the issues, and provide solutions. One of the first steps is to define the stakeholders and their needs. Some typical stakeholders and their needs are shown in ◘ Table 11.2.



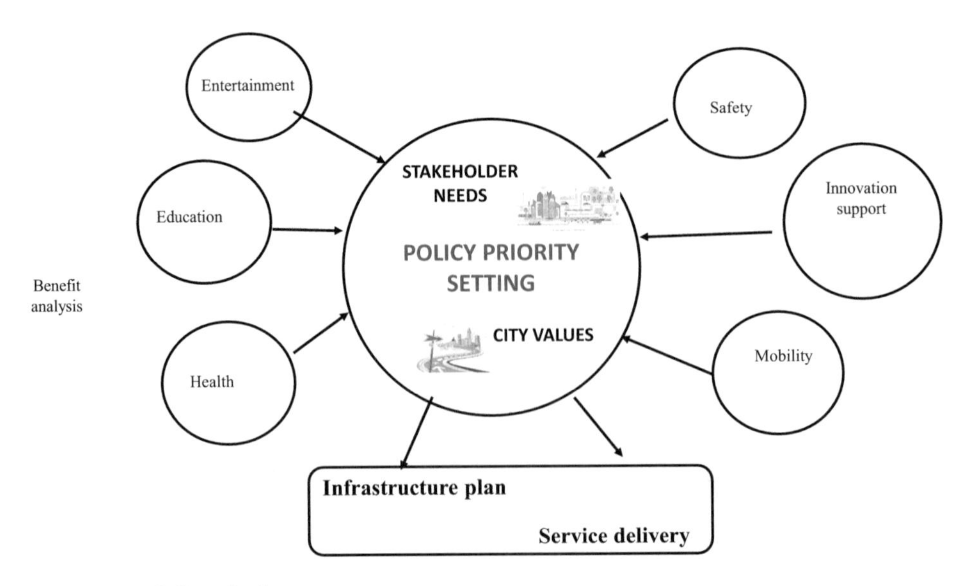


◘ **Fig. 11.4** Policy selection

**Table 11.2** Major needs of city stakeholders

| Stakeholder | Identified need |
|---|---|
| Resident | Ability to safely shop, move about and enjoy the environment |
| Tourist | Services to quickly get from place to place |
| Shop owner | Ability for customers to easily find the shop and display shop offerings |
| Police | Quick incident reports to enable quick response |
| Office workers | Quickly get to work |
| Transport workers | Manageable timetables with traffic conditions. Adaptable timetables |
| Government workers | Access to data to carry out planning |
| Families | Places to shop and find entertainment |
| Teachers and students | Availability of high-quality education |
| Health services | Ability to quickly access high quality medical services |

### 11.7.3 Collect Stories

Each city has stories about it, often based on people's experiences. However, city development most often is divided into the development of policies, sometimes linked rather than a single activity. Open innovation is increasingly used to collect citizens' ideas and needs.

### 11.7.4 Determine Issues and Their Causes

Processes used here are mainly consultative. There are council meetings, consultants preparing reports, community meetings, and sometimes even demonstrations. The methods described in ▶ Chap. 7 to create issue notes can be used here to record decisions. An example of an issue is shown in ◘ Table 11.3. It is also the case that city is different—they are of different sizes, have different demographics, are stronger in different areas of commerce.

We can then analyse the issues and identify causes as was done earlier using fishnet diagrams as shown in ◘ Fig. 11.5.

There is now a realization that even meeting city values is not enough. Increasingly, science has been calling for greater attention on taking care of the planet. Reducing the emissions of greenhouse gases becomes a priority to reduce possible dangers to our environment through climate change.

Here major conflicts are arising between governments and communities, with communities valuing environment, while others value commerce.

**Table 11.3** Example of issue note on transportation

| Checklist: | Smart cities |
|---|---|
| | Theme Smart City<br>Question: Are you happy with transportation? |
| Stake-holder | Stakeholder value judgement on question—As earlier in ▶ Chap. 7, we can summarize whether stakeholders agree (7) or not (1) |
| Shopper | 3 No, because it is getting hard to get shopping home if you come by public transport<br>6 There is a wide range of shops |
| Tourist | 4 Difficult to get to interesting sites |
| Local resident | 5 Takes a long time to get to work<br>2 Getting children to school takes too much time |

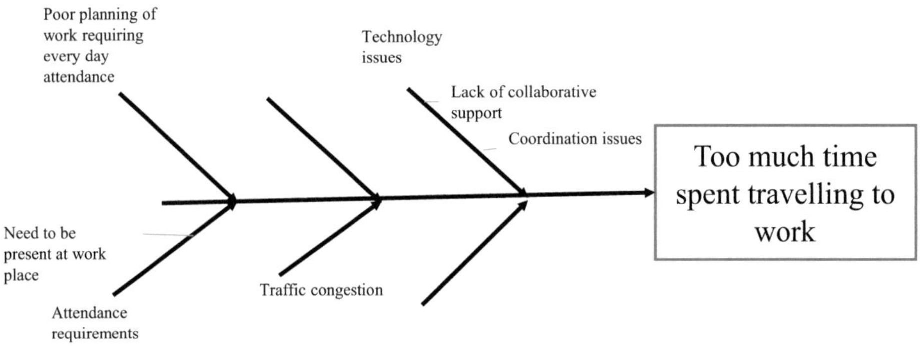

**Fig. 11.5** Example of issue analysis



We can then see the impact on city values. If the impact is high, then how will challenge be addressed? Examples include the following:

- Services needed to get people to work—combination of service and stakeholder.
- Services to get children to school—combination of service and stakeholder.
- Services to deliver goods—combination of client satisfaction and infrastructure.
- Services to help tourists and residents visit many sites.

### 11.7.5 Setting Policy Priorities

Priority setting is often a political issue. How to meet current needs but also have a vision for the future. Thus, values of city dwellers now form an important part of business model design. Such priorities can be:

- Reduce traffic congestion and improve mobility.
- Access to quality education.
- Provision of easily accessible health services.

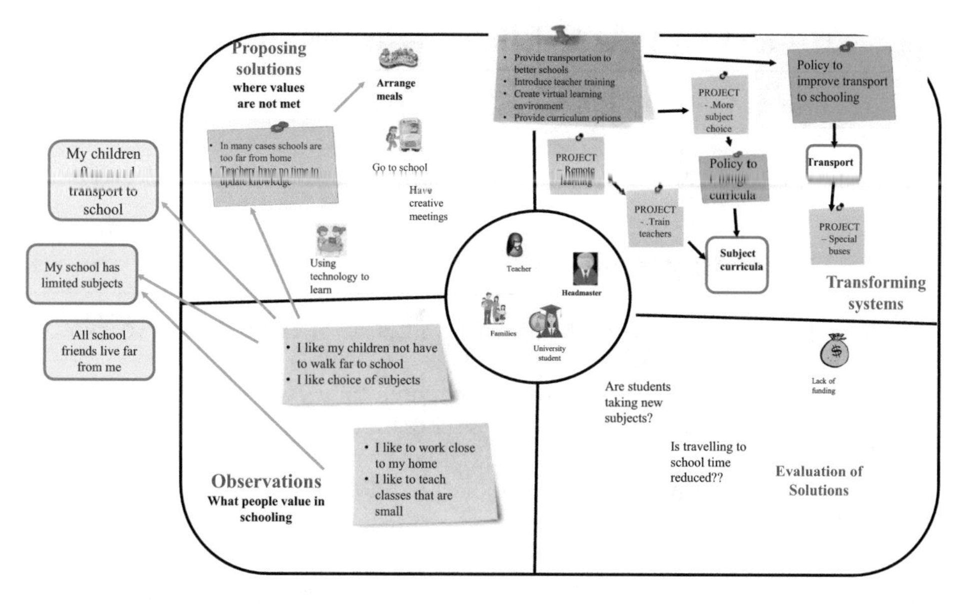


**Fig. 11.6** Education System Knowledge

## 11.8 Service Development

The next step is to provide services defined in the policy. It is possible to follow the same process as described in earlier chapters using the design circle shown in ► Fig. 5.2. The knowledge circle to develop services education policy is shown in ◘ Fig. 11.6. This circle only illustrates services for transportation, and those to realize greater curriculum choice.

The development of the knowledge proceeds in several steps and follows the knowledge circle.

- Step 1—Identify roles and place a symbol in the people circle.
- Step 2 (Q1)—Knowledge about the people and their values. How many of each role? Analysis and quantification of values. Knowledge of what they do.
- Step 3 (Q2)—what stakeholders see as issues.
- Step 4 (Q3)—identifying and implementing solutions.
- Step 5 (Q4)—Analysing outcomes.

◘ Figure 11.6 shows a few examples as post-it notes as data gathered is through the kinds of activities used in design thinking, often through brainstorming. For example:

- The blue notes show information on people values.
- The pink notes show how values are not being met.
- The orange notes show what can be done.
- The green notes show potential projects.

### 11.8.1 Q1—Identify How Stakeholders Are Affected by a Policy

As an example, Table 11.4 shows some of the stakeholders affected by education policies.

Again, such knowledge is often by discussion or brainstorming. For example:

### 11.8.2 Q2—Identify Issues

Issues on schools and education. Again, such values are often found by discussion or brainstorming. For example:

Examples: If parent value is "I like my children to walk to school" and schools are far away, then the gap is "in many cases schools are too far from home."

If a gap is "in many cases schools are too far from home" then ways to address may be "Provide transportation to schools." Table 11.5 illustrates one way to address this gap.

**Table 11.4** Some stakeholders impacted by education policy

| Stakeholder | What they do | What they value | Ways to satisfy value |
|---|---|---|---|
| Teachers | Deliver classes. | Good teaching materials. Well-designed classrooms | Raise quality of teaching |
| Children and parents | Attend school and learn | Easy access to schools and choice of subjects | Provide more choice in curriculum |
| School administrators | Allocate resources and oversee curricula. | Better learning outcomes | Define projects to improve learning |



**Table 11.5** An example of reasoning to develop education policy

| Issues (brief description of issues) | What needs to be addressed? | Define the data and knowledge needs to address problem | How will you get the needed data? | What do you see as a main challenge | What are the success measures? (Relate to values) |
|---|---|---|---|---|---|
| Issue: Buses go on regular routes that are not close to many students | Provide routes that go to schools safely | Get statistics on location of students that go to schools | Carry out surveys in schools | Getting resources to provide transportation | Reduced time for children to get to school |

### 11.8.3 Q3—Implementing Policies

The proposed solutions would require considerable consultation, brainstorming, and planning to develop a solution acceptable to most. Often the services are developed by businesses themselves working together with city authorities.

## 11.9 Infrastructure Policies

Policies can also focus on infrastructure. Transportation policy is developed to improve mobility in cities. Often these are political decisions made by city authorities. Infrastructure projects span application areas. Generally, large infrastructure projects, supported at government level, are beyond the scope of this book. These include building, hospitals, schools, and bridges.

Development of infrastructures is a study of its own and is not covered in detail in this book.

**Summary**
This chapter was a brief description of city design and ways cities are evaluated. It defined the trend to smart cities, the way to develop cities through policies that provide essential services to achieve smart goals. Policy briefs focus on some major theme, like health or education, and on infrastructure. Services are then provided to realize the policy goals.

**Exercises**

Select a city and find the policies for that city. Examples may be to compare the safety policies of cities like Singapore and London.

For your city, identify the policies that concern pandemic, especially COVID-19. What services support the policy?

## References

Appio, F. P., Lima, M., & Paroutis, S. (2019). Understanding smart cities: Innovation ecosystems, technological advancements, and social challenges. *Technological Forecasting and Social Change, 142*, 1–14.

Biswas, A. (2019). A framework to analyse inclusiveness in Urban policy. *Cities, 87*, 174–184.

Fennel, S., Kaur, P., Jhunjhunwals, A., Narayanan, D., Loyola, C., Begi, J., & Singh, Y. (2018). Examining linkages between smart villages and smart cities: Learning from rural youth accessing the internet in India. *Telecommunication Policy, 42*, 810–823.

Jacobides, M., Cennmo, C., & Gower, A. (2018). Towards a theory of ecosystems. *Strategic Management Journal, 39*(8), 2255–2276.

Lee, J. H., Hancock, M. G., & Hu, M.-C. (2014). Towards an effective framework for building smart cities: Lessons from Seoul and San Francisco. *Technological Forecasting and Social Change, 89*, 80–99.
Rebernik, N., Marusic, B., Bahillo, A., Osaba, B. (2019): "A four-dimensional model and combined methodologial to inclusive Urban Planning and design for ALL" *Sustainable Cities and Society, 44*, 195–214.
Zygiaris, S. (2013). Smart city reference model: Assisting planners to conceptualize the building of smart city innovation ecosystems. *Journal of the Knowledge Economy, 4*, 217–231. https://doi.org/10.1007/s13132-012-0089-4

## Further Reading

Camboin, G. F., Zawislak, P. A., & Pufal, N. A. (2019). Driving elements to make cities smarter: Evidence from European projects. *Technological Forecasting and Social Change, 142*, 154–167.
Letaifa, S. B. (2015). How to strategize smart cities: Revealing the SMART model. *Journal of Business Research, 68*, 1414–1419.
Schiavone, F., Paolone, F., & Mancini, D. (2019). Business model innovation for urban smartization. *Technological Forecasting and Social Change, 142*, 210–219.
Song, H., Srinivasan, R., Sookoor, T., & Jeschke, S. (Eds.). (2017). *Smart cities: Foundation, principles and application*. John Wiley.
Wang, H., & Yang, Y. (2019). Neighbourhood walkability: A review and bibliometric analysis. *Cities, 93*, 43–61.

### Web References (Last Accessed 12 March 2021)

Bristol: opendata.bristol.gov.uk/pages/air-quality-dashboard-new/map#air-quality-now
Helsinki: www.sustaineurope.com/smart-helsinki-20191025.html
OECD: http://oecd.org/greengrowth/compact-city-policies-9789264167865-en.html
San Francisco: sfcoit.org
Singapore: www.smartnation.gov.sg. Last accessed March 12, 2020.
Xiongan: A Smart City – Beijing Review (http://www.bjreview.com). http://www.bjreview.com/Nation/201804/t20180427_800128055.html

# Where to from Here?

Contents




I. T. Hawryszkiewycz, *Transforming Organizations in Disruptive Environments*,
https://doi.org/10.1007/978-981-16-1453-8_12

Transforming a business, city, or any other organization includes making changes to *the way organizations generate value* and *to the way people work* by the introduction of new processes and technologies.

At the time of writing, these questions are particularly relevant following the COVID-19 pandemic. The pandemic has shown the importance of value in transforming to new ways of working. Will organizations follow existing business models or adopt the new practices now emerging?

If anything, people often learn better ways to carry out their tasks and may adopt the new ways after the disaster. This chapter suggests some such ways and should be seen more as exploratory on what may happen in post-pandemic future. Innovations in technology emerge to support these new ways.

**Learning Objectives**

- Business models
- New ways people can work
- New ways for businesses to work

## 12.1 Introduction

Transforming any system requires changes to the way people work. Such change can come from technological innovation—as, for example, the arrival of train services changed the nature of society by widely increasing the ability to travel. Train services affected commerce by making it easier to deliver products quickly, while increasing tourism. Recently, the rapid technological innovation led to disruptions in many industries with new ways to conduct business emerging—online retail is the prime example. The advent of artificial intelligence is now beginning to have impact by providing a higher level of service.

By far the most major disruption, however, at the time of writing has been the COVID-19 pandemic. This chapter focuses on what is possible in a transformation, to what is seen as the post-COVID-19 world. The questions to be asked are the same as in any other transformation. These include how will businesses change? What is the future of work from home? How to continue to deliver the needed services safely? And how to eventually recover and deliver value to business stakeholders?

These questions must be answered in a situation where many jobs have been lost with people applying for emergency support in most countries. Hospitality and travel have been severely disrupted, while many other industries have responded by working from home. The nature of work and ways businesses work is clearly changing.



### 12.1.1 The New Normal

A new phrase has emerged—the new normal after COVID-19. Many writers see that a return to the way society worked in the year prior to the COVID-19 pandemic, generally taken to be the year 2019. Some see the "new normal" for work will be flexible, geographically dispersed, and offer a choice of a hybrid of digital and physical tools and spaces that are tailored to your role and your technology needs, preferences, and expectations. Work will be something you do, not somewhere you go.

The pandemic introduced more discussion on the future of work and how business will work in the future. There are now many questions being asked on the effect of the work. Work of course has been changing continually, recently with AI (artificial intelligence) seen as a great disruptor. In addition, the pandemic has introduced new possibilities, working remotely.

## 12.2 What Will Be the Industries?

There is a view emerging that the "post-COVID-19" world will also have many differences from the world before. There are many new ways that have been developed in the way people work, especially the growth of remote work, supported by quickly evolving technology. Acceptance of remote work will impact on business processes and business organizations.

*One obvious answer is health.* It is expected that COVID-19 will be active for some time. Contact tracing to identify clusters will be an on-going activity. Administrating the distribution of vaccines will be also important in early recovery.

*Greater emphasis on secure and reliable supply chains.* Business processes will need to become smarter and safer and greater attention will need to be paid between different steps of the business process. Journey maps and the associated touchpoint descriptions can be especially useful in such a design. Touchpoint design can focus on ensuring that all data needed in the touchpoint is provided and does not require additional contacts. It can also include elements like dashboards that provide complete knowledge of the progress of transactions.

*Growth of online shopping* resulting in robot operated warehouses.

*Managing data and deriving knowledge* will become increasingly important.

*Developing technologies* both to support contactless work practices and to support remote teamwork and collaboration in increasingly natural ways.

### 12.2.1 How Will Businesses Be Organized?

Here there are more questions than answers. Will there be a trend towards working from home? If so, will new processes be introduced for employees to work partially from home?

Will there be a trend to more individual work where people, rather than being employed by one business, will offer their skills to more than one business. Maybe instead of formal organizations, the trend will be to workforce ecosystems as suggested by Altman and others in the January 14, 2021 issue of Sloan Management Review, based on a trend to contract work, the evolving nature of work, trends to a diverse work force, and the growing complexity of workforce management in disruptive environments.

Use of scenario analysis to design such work ecosystems maybe the way to go.

## 12.3 How Will We Work?

*How will people work?* Increasingly, because of the pandemic, many workers are now finding working from home an attractive option with new technologies such as ZOOM making it possible to socialize at the same time, while exchanging and displaying documents.

The 2020 COVID-19 pandemic is expected to have a greater impact on the future of work. Working from home has now been shown to be possible, although some people question it on productivity grounds. Some people might question issues like catching nuances in meetings as important, but this may be balanced by the more focused approach in business activities. In many cases, tasks are better defined, business processes smoother. Can a balance be developed, where, perhaps with new emerging technologies, both can be achieved?

The possibility of networked organizations like those shown in ▶ Fig. 1.1, where virtual teams, which are composed of the most relevant experts, across the globe can address problems Arranging remote work includes both choosing technologies as well as rearranging what people do and how they work.

What then of the future after COVID-19? There will be both changes in the kind of jobs available and the way things are done. Will there be a trend to more contract employment to provide flexibility? The transformation must consider what is generally known as the post-COVID world—what will be the way we work. Nobody has an answer to this question. However, most accept that there will be more emphasis on working from home.

### 12.3.1 Digitally Enabled Work Arrangements



One comment here is that although many people have had positive experiences in using remote technologies in their work. It can, however, be argued, for example, that most exchange of knowledge has been of an explicit nature where people respond to requests. Meetings often focus on explicit data and discussions around a form or document. It can be argued that what is needed is.

- Seeing what you are doing in the global sense, so you see what you do in the context of the business.
- The kind of knowledge you, gather through gossip, chats in elevators, and random meeting that, in the aggregate, form organizational culture.

There is no doubt that we will see more technological innovation to enable people to immerse their work. Some possibilities come to mind. For example:

What about a display with people in your group "floating through space" inviting anyone for a quick chat? Individuals in your group or business unit can see who is ready for a chat and click on the person. Similarly, there can be an interface with an office layout that you wander through to see whose door is open and wander in for a quick chat.

## 12.4 What Will Be the New Business Models?

Porter's value chain is an early example where the idea of a value chain (Porter, 1980) was recognized to manage enterprises. A value chain focused on creating efficient internal processes that convert incoming resources into outgoing products utilizing the firm's resources.

Martin (2009) at the same time stressed the importance of innovation and the ability to quickly adapt to changing demand. His work focused on Design Thinking and the Idea of Open Innovation as practices by Procter and Gamble at the time.

The emphasis of managing resources has also been important since that time (Jay, 1991) emphasizing developing capabilities for resource management. At the same time, Nonaka and others (2014) defined the idea of a fractal organization, where resources were more focused on knowledge rather than physical.

Associated with the process, the idea of a business model emerged. The most popular was that of Osterwalder and Pigneur (2010) which identified the major business activities, some of which are illustrated in ◘ Table 12.1.

◘ **Table 12.1** Business model activities

| Business dimension | What are the measures | How does our proposal address the measures |
|---|---|---|
| Customer segment | People buying many bulky items | We are addressing people that make many purchases and find difficulty in carrying them. Mainly families and elderly |
| Customer channel | We will need to advertise the availability of the service | Providing services remotely<br>Contactless delivery |
| Customer relationship | Need to support communication between customers and deliverer | Find the best way to make it easier to arrange deliveries. |
| Key activities | Keeping track of deliveries.<br>Must ensure that platforms keep track of | Make remote work productive<br>Develop processes that are health safe<br>Avoid bureaucracy and simplify processes |
| Key resources | Need to purchase and develop technology, which will include developing the database. | We need to develop a platform for arranging deliveries of items<br>Cloud can be used to quickly raise customer base or new segments |
| Key partners | A new business unit to arrange deliveries to customers<br>Develop contracts with delivery people | We need someone to run the delivery system<br>How we transfer safely |

### 12.4.1 What About a Value/Capability Focus?

We can of course postulate on the emerging business models. Wang and Ahmed (2007) has argued over several years that a capability model is needed where the emphasis is on organizations developing capabilities that may involve more than one business part and, in this sense, develop the agility needed in the emerging world—a model that calls for continuous transformation.

Fractal organizations (Nonaka, 2014) is another proposal that suggests breaking up organizations into smaller business units that can be easily combined to address emerging challenges.

The idea of a fractal organization is attractive, especially given experiences in responding to natural disasters. We can also put it together with those that emphasize capability development.

### 12.4.2 How to Choose a Business Model?

We can of course talk about transforming in an ad-hoc way. However, in a more formal way, the change can be described as a change to the business model. A business model is how a particular business works. Changing the way a business works can look to a lot of discussion. A structured and systematic way is emerging to manage change.

Recently the term business model has become widely used to define the way businesses work. One of the most widely known models is, Osterwalder's model, which has nine dimensions that have been shown to cover all aspects of a business. Should transformations address these dimensions or adopt new models? But then, when you consider resilience and agility, which are increasingly important, how do you adapt any model to include them? Should there be a piecemeal approach, where you suggest ways to include them in each model dimension? Or should a set of capabilities be defined, and each dimension adjust to them?



It is always the case that any new service will impact the way the businesses will work. In the past, businesses were very structured—you get inputs, process them, and sell the outputs. This process was often referred to as a value chain, where the organization uses its expertise to add value to incoming. This simple model no longer applies because of increasing complexity resulting in a change to the business environment.

It is not the purpose of this book to introduce another business model. On the other hand, it is to draw from the ideas to adapt to the way the business works. So, this is seen here as an open question.

### 12.4.3 The Role of Technology

Technology will play an increasingly important role. Any transformation will use technology to make the new system work. The business model provides a framework for changing the business to include the new services.

Rapidly emerging technology and its dynamic nature itself adds to the possibility of developing dynamic capabilities, which enable rapid adaptation to disruption. An example may be AI supported by cloud capability (Djaja & Arial, 2015), which can be rapidly deployed to respond to disruptive events.

Technology can add value to all dimensions of the business model. It has been doing so for many years. However, there are many technologies that can now be used and choosing one can be challenging.

- What capabilities does big data provide?
- What capabilities can AI provide?

Technologies will no doubt play a role in any new business models or ways of working post recovery from the COVID-19 pandemic. Response has been swift in responding to the onset of the pandemic as witnessed on organizations such as ZOOM becoming almost a household name. As people adopt new practices, so technologies will be developed to support such practices. More interactive support for remote work will no doubt emerge. But what about artificial intelligence (AI) being used to manage contactless supply chains, or even design and adapt supply chains to meet a variety of user needs? These are all open questions today.

### 12.4.4 Will Open Innovation Become More Common?

Open innovation has been a practice used often in the past to collect ideas from different communities. One question that can be asked here is how to integrate open innovation into business processes. Such integration may become more important with the trend to remote work, where ideas or comments arising in casual conversations can be integrated into business processes.

Open innovation currently is most often outside the mainstream business process. There are many examples found in smart city development where ideas are posted on websites. These are then made available to the public but often the follow-up is not integrated into a business process.

## 12.5 A Review of This Book

It is perhaps fitting in this final part of this book to review its goals in terms of the material covered in this book. The focus has always been on values and the tools and methods with the goal of making informed decisions that add value to all stakeholders.

These values were described in ▶ Chaps. 3 and 4, but later expanded to include global values that society will increasingly face from disruptions emerging from climate change and increasingly growing cities.

This book also raised the importance of knowledge in ▶ Chap. 5 that showed relationships on the assumption that the more that is known, the better choice can be made in any transformation. Tools and methods to make such choices included a model, described in ▶ Chap. 6, that clearly showed the relationships found in any system so that the impact of any change on others are clearly apparent. ▶ Chapter 7 then described ways to analyse the system to identify issues and problems in attaining stakeholder values.

▶ Chapter 8 then described ways to address these problems in a systematic way where creativity and innovation is encouraged. At the same time, this book maintained awareness that transformations take place in a global environment that any transformation cannot ignore. It did so by stressing the importance of the triple bottom line.

Then in the final part of this book, ▶ Chaps. 9, 10, and 11, it expanded the importance of the triple bottom line into issues raised by climate change and the growth of cities indicating how tools were used in addressing global issues.

Summary
This short chapter explored possibilities for the future, especially focusing on new ways of working and ways the businesses will be organized. The emphasis was on working remotely and on business models that develop new business capabilities to quickly respond to emerging challenges, rather than focusing on business functions. It then concluded with a review of this book.

**Exercises**

You might think about how COVID-19 might affect work practices in your case studies and the trend to workforce ecosystems. For example, look at the food supply chain. Is it already on the way to becoming a workforce ecosystem? What about city development?

## References

Djaja, I., & Arial, M. (2015). The impact of dynamic information technology capability and strategic agility on business model innovation and firm performance in ICT firms. *Advanced Science Letters, 21*, 1225–1229.

Jay, B. (1991). Firm resources and sustained competitive advantage. *Journal of Management, 17*(1). https://doi.org/10.1177/014920639101700108

Martin, J. (2009). *The Design of Business*. Harvard Business Press.

Nonaka, I., Kodama, M., Hirose. A., Kohlbacher, F. (2014): Dynamic fractal organizations for promoting knowledge-based transformatio - A new paradigm for organizational theory. *European Management Journal 32*, 137–146.

Osterwalder, A., & Pigneur, Y. (2010). *Business model generation*. John Wylie and Sons.

Porter, M.E. (1980): Competitive Advantage. Free Press.

Wang, C. L., & Ahmed, P. K. (2007). Dynamic Capabilities: A review and Research Agenda. *International Journal of Management Reviews, 9*. https://doi.org/10.1111/j.1468-2370.2007.00201.x



## Further Reading

Altman, E. J., Kiron, D., Schwartz, J., & Jones, R. (2021). The future of work is through workforce ecosystems. *Sloan Management Review. January, 14*, 2021.

Kristenson, P., & Willel, L. (Eds.). (2019). *Service Innovation for Sustainable Business*. World Scientific Publishing.

Redante, R. C., Medeiros, J., Vidor, G., Cruz, C. M. L., & Ribero, J. L. D. (2019). Sustainable production and green product development: Using design thinking to promote stakeholders' engagement. *Sustainable Production and Consumption, 19*, 247–256.

Schoemaker, P. J. H. (1995). Scenario planning: A tool for strategic thinking. *MIT Sloan Management Review, 36*(2), 25–40.

Simms, M. (2019). *The future of work*. SAGE Publications.

Teece, D., Peteraf, M., & Leih, S. (2016). Dynamic capabilities and organizational agility: Risk, uncertainty, and strategy in the innovation economy. *California Management Review, 58*(4), 13–35.

# Supplementary Information




I. T. Hawryszkiewycz, *Transforming Organizations in Disruptive Environments*,
https://doi.org/10.1007/978-981-16-1453-8

# Index

## D



## E



## F



## G

## T



## U



## V



## W



## Z

Printed in the United States
by Baker & Taylor Publisher Services